A New Answer Book

By Mary Elting & Rose Wyler

Drawings by Rachel Folsom
Cartoons by Ferd Sondern

---- Consultants ----

Robert Moll
Assistant Professor
Dept. of Computer and Information Science
University of Massachusetts, Amherst

Eva-Lee Baird

Imperial Public Library
Imperial, Texas

GROSSET & DUNLAP
A FILMWAYS COMPANY
Publishers • New York

Imperial Public Library
Imperial, Texas

Library of Congress Catalog Card No.: 77-71531

ISBN: 0-448-12899-3 (Trade Edition)
ISBN: 0-448-13418-7 (Library Edition)

Copyright © 1977 by Mary Elting.
All rights reserved under International and Pan-American Copyright Conventions.
Published simultaneously in Canada. Printed in the United States of America.

Contents

CHAPTER   PAGE

I: ARE YOU LUCKY?   7
Do people ever have lucky streaks?...Are there any lucky numbers?...Is 13 an unlucky number?...What is a fifty-fifty chance?...What is a long shot?...Would you believe it?...Does it pay to keep on buying tickets in a state lottery?...What is the law of averages?

II: MONEY, MONEY, MONEY   13
How could you get rich quick?...Can you make a fortune with chain letters?...Why don't all people use the same kind of money?...Where does the word *dollar* come from?...Where did the dollar sign come from?...When is a penny not a penny?...Where did the pound sign come from?...Where do "heads" and "tails" come from?...How did banks get started?...What is an income tax?...What is a minimum wage?...Is a loan shark an animal?...What does it mean to go bankrupt?...What is a profit?...How does a bank make money by giving you money?...Can making money cost too much?...What does a million dollars look like?

III: THE METS, THE JETS AND THE NETS   27
What was the four-minute mile?...What is a metric mile?...What is a batting average?...Why is a track field shaped the way it is?...Why is a baseball field called a diamond?...What is a Quarter Horse?...Why are professional football players paid so much?...What does love in tennis mean?...Which swimming stroke is the fastest?

IV: FORTUNETELLING AND LOOKING INTO THE FUTURE   33
Were you born under a lucky star?...How did astrology begin?...Is a comet an evil omen?...What do people mean when they say "I've got your number"?...How magical are magic squares?...Can a mind reader really read your mind?...What is ESP?...Can cards tell your future?...Are there any real ways to predict the future?...What is the census?...How do public opinion polls work?

V: NUMBERS, THE GREAT INVENTION   43

Why do we call numbers a great invention?...How did names for numbers get started?...Where did our numerals come from?...How did the Romans add with Roman numerals?...Which came first: zero or one?...Who were the zero freaks?...Can you count without numbers?...Is zero the same as nothing?...Can anything be less than nothing?...What is the smallest number in the world?...Did people ever count on their toes?...Do you remember how you learned to count?...Why do astronauts use "count-down" instead of "count-up"?...Can any animals count?...Why do we count eggs by the dozen?...Who invented fractions?...What are decimals?...Who invented decimals?...What is a round number?...What is a square number?...Are there any triangular numbers?...Does 1 + 1 always equal 2?...Do all people count the same way?

VI: FIGURING IT OUT   57

How can you add without writing down numbers?...Which is faster: an abacus or an adding machine?...Where did the signs for "plus" and "minus" come from?...Where did the sign for "multiply" come from?...Where did the sign for "divide" come from?...Where did the sign for "equal" come from?...Can you multiply just by adding?...How can 10 + 5 = 3?...What is the difference between arithmetic and algebra?...What does x mean?...Who invented the computer?...What is a computer program?...What is a bug in a computer program?...Can a computer remember all the numbers in a phone book?...How does a computer's memory work?...Can a machine play chess?...Can a computer make you richer?...How does a pocket calculator work?...Is there any difference between a calculator and a computer?...Can you make a computer cheat?...Can computers think?...What has digits but no hands?...What was the first digital computer?...Why do the numbers on a pocket calculator look so strange?

VII: SO BIG, SO SMALL   73

When was a foot not a foot?...How can the ocean's depth be measured?...How far is it to the bottom of the sea?...Can a fat man reduce by going to the top of a mountain?...Which weighs more: a pound of feathers or a pound of gold?...What did these measures measure?...And what about these?...What is a hairbreadth escape?...Why didn't the United States adopt the metric system when other countries did?...How do we know the length of a meter?...Why do thermometers have different scales?...How do we know how many calories are in a hamburger?

VIII: ROUND AND SQUARE, THICK AND THIN   81

Could a bird sit on a square egg?...How far away is the horizon?...Why does the earth look flat when it is really round?...Why is pi such a strange number?...Are there any lucky shapes?...How many of these shapes can you name?...Why are dice always cube-shaped?...What is a square deal?...Can a dog draw a circle?...Why do starfish have five arms?...Why are there right angles but no left angles?...Do stars really have five points?...Why are bananas curved?

IX: WHAT IS IT WORTH?   89

What does unit price mean?...Is anything cheaper by the dozen?...Why is 13 a baker's dozen?...What do numbers on a milk carton mean?...How do you read an electric meter?...What is a sales tax?...What is pin money?...What is a budget?...What is double-digit inflation?...Why does a phone bill have holes in it?...How does a credit card work?

X: ON THE ROAD 95
How can you figure miles per gallon?...Why can you go farther on a gallon of gas in Canada than in the United States?...How far can a car travel on a gallon of gas?...Is a speedometer the same as an odometer?...Why are there both letters and numbers on some license plates?...How do you read the mileage table on a road map?...What does a "measured mile" mean?...When was a mile not a mile?

XI: TIME AND SPACE 99
Why does the calendar have twelve months?...How do we know the length of the year?...Why are there seven days in a week?...Why isn't Easter on the same date every year?...Why do Jewish holidays come on different dates from year to year?...Why are there 24 hours in a day?...Why are there 60 minutes in an hour?...What is a great-circle route?...What is jet lag?...How can you lose a day?...What is a light year?...What do B.C., A.D., and B.P. mean?

XII: NUMBERS IN YOUR LIFE 107
Why do we need telephone numbers?...Why do telephone dials have both numbers and letters?...Can you call a friend without using a phone number?...Why do we need area code numbers?...How can a family have 3.8 people?...Do you grow the same number of inches every year?...When people say you are above average, what do they mean?...Can a doctor really tell how many red blood cells you have?...Why do most library books have numbers?...Why don't all library books have numbers?...Why do we have ZIP codes?

XIII: WONDERFUL AND FAR OUT 115
Would numbers exist if no one had ever thought of them?...What is the biggest number you can write with three digits?...What is infinity?...Why are some numbers called imaginary numbers?...Can a piece of paper have only one side? ...How can you make something twice as big by cutting it in two?...Why is a Halloween mask like a pretzel?..."How many bulls' tails are needed to reach the moon?"..."How many black beans does it take to make three white beans?"... Can you learn to be a lightning calculator?...Which is shorter: an uphill mile or a downhill mile?...What is a time capsule?...Can we signal to people in other worlds?...What is the fourth dimension?...And now, our final question: Where did all these answers come from?

INDEX 126

Chapter I

Are You Lucky?

1. Do people ever have lucky streaks?
2. Are there any lucky numbers?
3. Is 13 an unlucky number?
4. What is a fifty-fifty chance?
5. What is a long shot?
6. Would you believe it?
7. Does it pay to keep on buying tickets in a state lottery?
8. What is the law of averages?

DO PEOPLE EVER HAVE LUCKY STREAKS?

Bingo! You've done it again! Your bingo card is a winner for the eleventh time in a row. You are having an unbelievably lucky streak.

Lucky streaks do not happen very often, but they do happen. The same is true of unlucky streaks — that is, losing streaks. They happen, and they, too, are rare.

Luck is just a matter of chance. It is something no one can control or develop.

When your basketball team wins nine games in a row, is that a lucky streak? No. Basketball is a game of skill, not a game of chance. Your team does not win basketball games by luck. It wins because your players are better.

ARE THERE ANY LUCKY NUMBERS?

Imagine that you lived in very ancient times when people were just finding out how to use numbers. One day you made an exciting discovery — many important things came in pairs! Arms, feet, legs, eyes, hands, man and woman, day and night. Two, it seemed, was a very important number.

Good things also came in threes: father, mother, child; water, earth, sky. Other discoveries followed. A hand had five fingers. The head had seven openings: two eyes, two nostrils, two ears, and one mouth. Certain heavenly bodies moved across the sky in regular ways at regular times. And there were seven of them: the sun, the moon, and the five planets that are visible without a telescope.

Every number, long ago, meant something special. A person who understood numbers seemed to have special powers. In fact, numbers themselves seemed powerful, almost holy. Perhaps they could even bring good luck.

In many places seven was considered the luckiest number of all. Japanese people believed there were Seven Gods of Good Luck who were supposed to sail treasure ships into port on New Year's Eve. Europeans thought that extra good fortune would come to the seventh son of a man who was a seventh son.

Ancient Romans thought that ten was lucky because the gods had given human beings ten fingers to count with.

When people learned how to multiply and divide, six became a great wonder. They found that it could be divided by 1, 2, and 3, and that 1 + 2 + 3 made 6. This seemed so remarkable that six was called a *perfect number*. And, of course, a perfect number had to be lucky.

People still want numbers to be lucky. One person believes that he will win if his number in a lottery ends with the same number as the day he was born. Another believes she will have good luck if she finds a four-leaf clover. There are hundreds of such beliefs, but scientists are unable to find proof that any number has an effect on luck.

The "Thirteen Goddess," an ancient Zapotec Indian statue. The number 13 on her breast is written with two bars and three dots.

How many 13's can you find on the back of a one-dollar bill?

IS 13 AN UNLUCKY NUMBER?

In many countries of the world people think the number 13 is unlucky. Someone you know can probably tell you about a terrible thing that happened on the 13th day of the month. Of course, just as many unfortunate things happen on other days, but people often forget this.

There was a time when businessmen would not rent an office on the 13th floor of a building, and families did not like to live in apartments on the 13th floor. So the owners of buildings simply numbered the higher floors 10, 11, 12, 14, 15, and so on. Nothing bad happened to people on floor number 14, which was really number 13, so now the 13th floors usually have that number.

In China and Egypt people once thought the number 13 brought good fortune. In Belgium, women wore good-luck charms formed in the shape of 13. No one knows why that particular number got a good reputation in some places and a bad reputation in others.

In the United States, 13 isn't always considered unlucky — at least, not by the people who use dollar bills. On the back of these bills is a picture of a pyramid with 13 steps. Opposite it is an eagle clutching 13 arrows and an olive branch with 13 leaves and 13 berries. The shield in front of the eagle has 13 stripes, one for each of the 13 original states.

WHAT IS A FIFTY-FIFTY CHANCE?

John holds out two fists. A piece of candy is hidden in one of them. "Guess which one," he says. "You have a fifty-fifty chance of being right." And that means the candy is just as likely to be in one fist as in the other. The chance of guessing right equals the chance of guessing wrong.

But do you guess right? No? Well, your chance of guessing wrong was also fifty-fifty.

Now John sets five cups in a row, all upside-down. He asks you not to look as he puts a piece of candy under one of them. "Ready," he says. "See if you can pick out an *empty* cup."

This time your chance of guessing right is much better than fifty-fifty. You have four chances out of five of picking the empty cup.

What do you pick? The cup with the candy under it! You chose the wrong one, although there was less than a fifty-fifty chance of doing so.

What do you do with the candy? Eat it, of course. But first you divide it into two equal pieces and share it, fifty-fifty.

WHAT IS A LONG SHOT?

Sometimes it happens: you flip ten coins, and ten "heads" turn up. Yet there is only about one chance in a thousand of getting ten in a row. Anything as rare as that deserves a special name, and it has one. It is called a *long shot*.

The name is old. About seven centuries ago, nearly every Englishman knew how to use a bow and arrow. When a town held an archery contest, men would take aim at a target almost a quarter of a mile away. Hitting the target was known as making a long shot. Few hits were made, so each one seemed very remarkable. Soon people were using the words *long shot* for any result that was surprising and unlikely — just as we do today.

WOULD YOU BELIEVE IT?

Guests were once afraid to eat dinner at a table where there were 13 people. This fear of thirteen has a special long name: *triskaidekaphobia!*

Certain ancient Greeks hated the number seventeen. They said it was ugly because it came between two beautiful numbers: sixteen and eighteen. Sixteen they liked because it is four times four, and four represented a square. Eighteen was two nines, and nine was the lucky number three multiplied by itself.

Two curious numbers are 220 and 284. If you add together all the numbers that can be divided into 220, you get 284. Add all the numbers that can be divided into 284 and you get 220. Mathematicians called 220 and 284 *amicable numbers*. That meant they were agreeable, friendly.

Could these special numbers be used to guarantee that two people would always be friendly? Girls and boys used to think so: before getting married they wrote 220 and 284 on little pills. The boy swallowed one and the girl swallowed the other. And in so doing they were supposed to live happily ever after.

DOES IT PAY TO KEEP ON BUYING TICKETS IN A STATE LOTTERY?

The prizes in state lotteries are big — very big. In some places it's possible to win a million dollars with a single 50¢ ticket. It would seem that such lotteries give people a fantastic chance to get rich quick. But there's a catch.

The chance of winning is really very small. If there is one prize and 500,000 tickets are sold, a person with one ticket has just one chance out of 500,000 of winning. That's the same as having 499,999 chances of losing. Buying two tickets does not improve the chances very much. Two tickets give a buyer two chances out of 500,000. Three tickets give three chances, and so on.

If there is a yearly prize, and if you buy one ticket a week, you still have only 52 chances out of 500,000 to win.

What happens to the money collected from the sale of tickets? Do the winners get all of it? No! In the United States the prize winners often get only $40 out of every $100 paid for tickets. About $10, and sometimes even more than that, is used to pay the cost of running the lottery. The rest goes to the government to use for schools and other services.

Some people think lotteries are not a good way to raise money. The regular players are usually poor, and very few of them ever win back the money they spend on tickets.

WHAT IS THE LAW OF AVERAGES?

Suppose you have three bags of pennies. The first bag holds 10 pennies, the second bag holds 100 pennies, and the third bag holds 1,000 pennies.

Spill the pennies in the first bag on the floor. How many heads turn up, and how many tails? Count the number of heads. Let's say there are 6 of them. Since there are 10 coins in all, 6 out of 10, or 60%, are heads.

Now spill the second bag, the one with 100 pennies, and count the heads. This time, suppose there are 43. This means that 43 out of 100, or 43%, are heads.

Empty the last bag. Checking 1,000 pennies may take a while, but when you are done, let's say you've found 533 heads. This is 53.3% of the 1,000 coins on the floor.

When you spill a bag of pennies, the "law of averages" tells you what *should* happen. This law is a mathematical rule. It says that you should get close to 50% heads. It also says that the bigger the bag (with more coins), the closer the number of heads should be to 50%. So if a dump-truck full of 100,000 pennies shows up in your driveway, you may as well tell the driver not to unload. The law of averages tells you that you would probably find close to 50,000 heads if he spilled them all out!

Chapter II

Money, Money, Money

1. *How could you get rich quick?*
2. *Can you make a fortune with chain letters?*
3. *Why don't all people use the same kind of money?*
4. *Where does the word* dollar *come from?*
5. *Where did the dollar sign come from?*
6. *When is a penny not a penny?*
7. *Where did the pound sign come from?*
8. *Where do "heads" and "tails" come from?*
9. *How did banks get started?*
10. *What is an income tax?*
11. *What is a minimum wage?*
12. *Is a loan shark an animal?*
13. *What does it mean to go bankrupt?*
14. *What is a profit?*
15. *How does a bank make money by giving you money?*
16. *Can making money cost too much?*
17. *What does a million dollars look like?*

HOW COULD YOU GET RICH QUICK?

Would you rather be paid $10 a week, or a penny for the first day and double your pay each day after that?

If you get $10 a week, you will have $10 at the end of one week and $40 after four weeks.

If you start with a penny and double your pay each day, you will finish the first week with $1.27. Here's why. The first day you get 1¢. The second day your pay is doubled, so you get 2¢. It is doubled again on the third day, so you get 4¢. The next day you get 8¢, then 16¢, 32¢, and 64¢. That ends the first week, and you have

1¢ + 2¢ + 4¢ + 8¢ + 16¢ + 32¢ + 64¢ = $1.27

Of course, $1.27 is a lot less than $10. But what happens in the second week? You start with 64¢ doubled, or $1.28. Doubling your pay each day, you get $2.56, $5.12, $10.24, $20.48, $40.96, and $81.92. When you add up your money for the first two weeks, you will find that you have $163.83. Pretty soon you'll be rich!

After three weeks, you have $20,971.51. One week later you get your first check for a million dollars — $1,342,177.27, to be exact. And 23 days after that, you get a check for over 11 trillion dollars — *that's almost all the money in the world!*

CAN YOU MAKE A FORTUNE WITH CHAIN LETTERS?

One day in 1935 strange letters began coming to people in Denver, Colorado. They said something like this:

You can make $1,562.50 in a week!

Follow these instructions *TODAY:*
1) Send 10¢ to the first name on the list below.
2) Make 5 copies of this letter, but omit the first name on the list and put your name and address in fifth place.
3) Send your copies to 5 friends. Be sure they are people who will follow the instructions.

Six days from now your name will be at the top of the list in 15,625 letters. That means 15,625 people will send you dimes. In the next mail you will receive $1,562.50!

John Able, 9 Federal Blvd., Denver, Colo.
Mary Baker, 10 Grove St., Denver, Colo.
Sue Charles, 11 Harris St., Denver, Colo.
George Dee, 12 Irving Pl., Denver, Colo.
Jane Echo, 13 Jackson St., Denver, Colo.

DO NOT BREAK THIS CHAIN

The chain-letter idea sounded wonderful. It would cost just a dime, plus 2¢ on each letter for postage. (Letters needed only a 2¢ stamp in 1935.) Before long, hundreds, then thousands of letters began to pour into post offices. One post office in Denver handled 95,000 letters in a day, and the idea spread to other cities.

People got so excited that they sometimes put dimes into envelopes and forgot to address them. Or they copied addresses incorrectly. Soon the Denver post offices had 100,000 envelopes containing dimes that could not be delivered.

No one knows who started the chain letter epidemic. It lasted a few months and then faded away. By that time post offices all over the country had piles of letters with wrong addresses or with no addresses — 3,000,000 of them altogether!

But did anyone make a fortune? The scheme would work only if everyone who got a letter sent a dime to the first person on the list and then sent copies of the letter to 5 friends. Many people would not bother. But let's suppose that you got the letter and mailed copies to 5 friends who received them on Monday. Then each of them sent out 5 letters, a total of 25. The 25 who received them on Tuesday would send out 5 each, or 125. On Wednesday it would be 125 × 5, making 625. On Thursday, 625 × 5, making 3,125. On Friday, 3,125 × 5, making 15,625. And on Saturday 15,625 people would send you dimes, because your name would now be at the top of the list. That many dimes is $1,562.50.

Of course, if everybody continued to follow instructions, the chain would go on. At the end of the second week more than 200,000,000 letters would have been mailed — more than all the men, women and children in the United States at the time the chain-letter idea started!

Chain letters never made anyone really rich. The people in the first chains in Denver probably did get a good many dimes, because the idea was new. Perhaps someone even got $1,562.50. But when friends sent letters to friends who had already received letters, the names began to overlap. One man got more than 2,000 letters. He finally put an ad in the newspaper saying he would not send out any dimes. Many people who sent out dimes did not get any back. So they ended up poorer instead of richer.

What became of the dimes in letters that could not be delivered? They went to a Dead Letter Office where the United States' Postal Service holds mail with a wrong address or no address at all. Letters containing money are kept there for a while. If nobody claims them, the money is used to help pay post office expenses.

WHY DON'T ALL PEOPLE USE THE SAME KIND OF MONEY?

Every kind of money in the world has its own story. But all the stories start in the same way: one person wanted something another person had. So person Number 1 offered Number 2 an item that both of them valued. And a deal was made. In a great many parts of the world people valued shells, so shells became their first money. In other places, cattle had such great value that they could be used to buy almost everything else.

Many thousand years ago, groups of people in different parts of the world began to develop their own ideas about what could be exchanged for what. As a rule, ancient money was something to wear, eat, drink, use as a decoration — or boast about. Often it was something scarce or hard to get. Salt, for example. Salt hunger bothered many people after hunters learned to be farmers. Hunters ate meat, which contains salt, but farmers ate grains, which do not. And so cakes of salt were used as money by many farming people.

Grain itself was money in ancient Egypt. In Asia, tea mixed with sawdust was molded into small blocks and used for buying anything from camels to fishhooks. Other unusual money included woodpecker scalps, cocoa beans, animal-tooth necklaces, little bells. Until recent times big chunks of stone with holes in the middle were money on the Island of Yap.

Gold, one of the scarcest things, became money in ancient Egypt in a strange way. For a long time the Egyptians had no gold of their own. Instead, they stole the gold beads that their neighbors wore. The Egyptians ground up the beads and used the gold dust as money, just the way prospectors did during the Gold Rush days in North America.

At last somebody hit on the idea of making gold and silver coins. The idea spread from one place to another in Africa, Asia, and Europe, and then across the oceans. Each country had its own special coins. So did some important cities.

Every country in the world now has laws saying what kind of money is to be used and how it is to be manufactured. If you travel from country to country, you can exchange one kind of money for another. But each government has rules about the exchange. And just as most people want a government of their own, they want their own kind of money, too.

Odd-shaped coins are useful among people who can't read. It is easy to remember a shape, even if you can't read the numbers on a coin.

For about a thousand years people on the Island of Yap used money called *fei* made from chunks of stone. A twenty-inch *fei* would buy a pig.

WHERE DOES THE WORD *DOLLAR* COME FROM?

Five hundred years ago the King of Bohemia was the only ruler in Europe who could get almost as many silver coins as he wanted. The silver came from his mines in a place called Joachimsthal. At first these coins were called *Joachimsthaler*. Then the name was shortened to *thaler* or *taler*. Bohemian *talers* were used in other countries that had few coins of their own. Later, when some of those countries adopted new money, people still called certain coins *talers* or *dalers* or *dolars* or *dollars*.

WHERE DID THE DOLLAR SIGN COME FROM?

Before the American Revolution, most people in the thirteen English colonies had English money. There were also Spanish colonies in North America, and they had Spanish money. But coins from both Spain and England could be used almost anywhere in North America.

"Piece of eight," or Spanish dollar, the ancestor of the United States dollar.

Metal bells and throwing knife used as money in Africa.

English-speaking merchants actually preferred Spanish coins because it was easy to do business with them. A hundred small Spanish coins were equal to one large one, and merchants had no trouble multiplying or dividing by 100. But English money was complicated. Twelve of the small English coins equalled the next larger one, and twenty of those equalled the next larger. It was a nuisance to keep figuring with twelve and twenty.

One kind of Spanish coin was called a *peso*. Whenever a person owed a store 5 pesos and then paid the bill, the storekeeper usually did not take time to spell out "5 pesos." Instead he wrote 5 p^s. Or, if he had sloppy handwriting, he often just made squiggles like these

As you can see, one of these careless peso signs happened to look better than others. People began using it and making it neatly, like this: $, or like this: $.

After the American Revolution, the new United States needed its own kind of money. Since the Spanish system was convenient, the United States government decided to copy it. Now 100 coins, called cents in the new United States money, equalled a new coin called a dollar. And the familiar Spanish peso sign became the dollar sign.

In Mexico, *peso* is the name for the main unit of money. The Mexican peso sign looks like the dollar sign that has only one bar through the S, like this: $.

For a while after the American Revolution people in Canada used the complicated English "twelve-and-twenty" coin system. But then they found that it would be easier to trade with the United States if they had similar money. So they adopted a Canadian system of dollars and cents, and their dollar sign is the same as the United States dollar sign.

Malaysia, Australia, and Singapore also use this same sign for their dollars.

WHEN IS A PENNY NOT A PENNY?

In the United States, a penny is not a penny when it is in the mint. That is the factory where money is made. In the official language of the United States government there is no such coin as a penny. Its proper name is cent, and 100 cents make a dollar. In England, penny is an official name, and 100 pennies make a pound. English people also use the word *pence* for pennies.

Penny is probably the oldest coin name in the English language. A thousand years ago people were already using it, probably because the coin was shaped like a pan, and the ancient word for pan was pronounced *pah-neh*. In early days a penny was made of silver. In the year 1266 King Henry III said it had to weigh exactly the same as 32 grains of dry wheat. It was worth much more than the modern penny. You could buy a pig with just one silver penny.

"Heads and tails," Greek coins 2,300 years old.

WHERE DID THE POUND SIGN COME FROM?

In the days before there were banks, almost everybody who had money kept it in his own safe place. And he didn't count it — he weighed it. The ancient Romans used money-weighing scales which were called *libra,* and when a pile of silver coins was weighed, they said "*libra pondo.*"

After a time, Roman ideas and words traveled north to England. But, so far from home, the words did not remain quite the same. *Pondo* became *pound.* A pound came to be a unit of measure for money and also for ordinary things such as wheat or fish. Of course, a pound of silver coins was worth much more than a pound of fish.

Ideas about money changed slowly. Kings began to have coins made of silver mixed with a much cheaper metal. Now a pound of coins wasn't the same as a pound of silver. People stopped weighing their money, but they kept the word pound. And an English pound is still a money unit.

Meantime, the old Roman word for scales changed, too. *Libra* was shortened to lb. Written in small letters like that, it meant pounds of ordinary things — 2 lb. butter, for example. But when a bookkeeper wrote down a sum of money, small letters were not important enough. He made a big capital L, then added a little extra flourish, making £.

WHERE DO "HEADS" AND "TAILS" COME FROM?

For at least 2,500 years people have used coins for buying and selling. In ancient times most countries were ruled by kings who made decisions about money. Usually a king had his own picture stamped on one side of a coin. And so, even in the days of ancient Rome, one side of the coin was known as the head.

If one side was a head, then, people said, the other side must be the tail. And they said "tails," even though few coins ever had a tail on the other side. Some early coins did have designs or pictures of animals on the tail side. But on many ancient coins, the tail side was left entirely blank.

Shipping and trade in Germany, five hundred years ago. From an old manuscript.

HOW DID BANKS GET STARTED?

The first banks in the world were warehouses. They belonged to kings and religious leaders who collected gifts and taxes from ordinary people. After a while the rulers and priests had so much treasure stored away they did not know what to do with it. Eventually they let farmers borrow some to buy seeds and they gave loans to jewelry makers and other craftsmen. Of course, the borrowers had to pay back the money, plus a little extra each time. And so the kings and priests got even richer than before. That was more than five thousand years ago in a land called Sumer.

Time passed, and priests in many places stored the valuables they collected in strong buildings called treasuries. These, too, were banks. Treasuries were safer than warehouses, but when there were wars, treasures such as gold and silver were often stolen by raiding armies. No one knows where all the stolen treasure went. Some of it was lost. A great deal of it was buried in boxes and jars for safekeeping, and the owners never came back to get it. Once in a while these secret ancient treasure hoards are still discovered.

A money changer in ancient times. From a woodcut by Hans Weiditz.

For hundreds of years in Europe there was so much warfare and misery that ordinary people had almost no money. They certainly didn't need banks. Kings and priests still collected taxes, of course. Princes and lords managed to become rich, too, and some of them kept their gold and silver in a place called the mint. That was a kind of workshop where coins were made. But the mint was not always safe, because a king might just take the gold for himself if he wanted to pay for a war or for any other expensive habit he had.

The rich lords in England wanted a safe place for their wealth, and they found one. At that time they used cups and plates and bowls made mostly of gold and silver. The goldsmiths who made the metal dishes had storerooms for their supplies of precious material. They were usually honest men, and so rich people began to keep money and valuables in goldsmiths' shops. Later somebody had the idea of writing checks instead of using coins. The goldsmiths who cashed checks often did other money business, too, and they became bankers.

About 700 years ago men began to build bigger ships, and traders in Europe began to do more business. Many of the ships came regularly to the city of Venice in Italy. They brought spices and gems, cloth, furs, rugs from far-off countries. Since each country had its own special coins, a trader might need a dozen different kinds of money on a long trip. When he reached Venice he often wanted to exchange one kind for another.

Some coins were very valuable, while others weren't worth much. Only experts could figure out how to exchange one kind of coin for another. The expert money-changers in Venice kept boxes and baskets of many different coins on shelves and counters in their shops. The counters were called *bancas*. Merchants who visited the shops always had to pay a fee when they traded one kind of coin for another, and clever money-changers often grew rich. Like the goldsmiths, they began lending money and making other business deals. To traders, the *banca* in a money shop was a place of great importance. And that is where the word *bank* comes from.

Money-changers weren't always honest. Sometimes they cheated traders. When that happened, a trader had the right to break up the counter in the dishonest money-changer's shop and put him out of business. In the Italian language a broken counter was a *banca rotta*. The two words became one word in English — *bankrupt*. And now *anyone* who has to go out of business may become bankrupt.

WHAT IS AN INCOME TAX?

A tax is money that people pay to their government. The government then uses the money in many ways. Taxes pay for roads, parks, scientific expeditions, school lunches, wars, and thousands of other things.

Should everyone be taxed? How much? What is fair? Many people think an income tax is the fairest kind. Those who earn very little money should pay the smallest taxes. They need most — or even all — of their income to buy food and other necessary things. People who have a very large income should pay the most, because they use only a small part of their money for necessities.

At one time or another many kinds of things have been taxed. In Italy there once was a tax on the windows in a house. At one time in the United States a man had to pay a special tax if he owned a gold watch. It is only in modern times that we have had income taxes.

In ancient Egypt this tax collector was paid in geese, not in money.

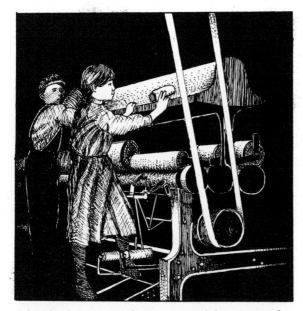

A hundred years ago there was no minimum wage for children who worked in American factories.

WHAT IS A MINIMUM WAGE?

About a hundred years ago school children in the United States found this problem in their math books:

> What sum can be saved in 40 years by rising 45 minutes earlier, 300 days of each year, if an hour is worth 15 cents?
>
> Answer: $1,350

At that time many people were paid only 15 cents an hour. Sometimes their wages were even smaller. At last working men and women began to complain. Many of them joined together in labor unions, and the unions helped to get them better wages. Now, in some countries, there are special laws about wages for certain kinds of work. The laws set an amount that is the *least* money an employer may pay. Another word for least is *minimum*. And so a minimum wage is the smallest wage allowed by law.

22

IS A LOAN SHARK AN ANIMAL?

A loan shark walks on two legs and doesn't live in the ocean. Maybe you have seen one depicted on television. A loan shark is a person who lends money to people and then beats them up if they can't pay the money back. Usually the borrowers are in trouble and can't get money from friends or from a bank.

The loan shark knows that people who borrow from him are desperate. So he makes them promise to pay back two or three or four times as much as they would pay if they borrowed from a bank. It is against the law in most countries to charge that much for a loan. And, of course, it is illegal for a loan shark to harm borrowers if they can't pay on time.

Most people are afraid of both loan sharks and real sharks. When English sailors first caught a real shark more than three hundred years ago they said the shark was so mean it could kill a swimmer. Ever since then, "shark" has been a slang word for a mean person who hurts or cheats others.

WHAT DOES IT MEAN TO GO BANKRUPT?

Suppose you want to start a business of your own. You decide to have an ice cream store. You buy tables, chairs, freezers, dishes, mixes — everything. Lots of customers come.

But one day the electric power goes off. The freezer can't freeze. All the ice cream melts, and you have to throw it out. You make more when the electricity goes on. Now you have plenty of customers. But some of them begin roughhousing and you have to throw *them* out. The others get angry and don't pay. Then a car runs up on the sidewalk and breaks your store window, so you have to buy a new one. Then . . .

By now you are spending money faster than you get it from customers. You can't pay the rent, or the electric bill, or any other bill. You are in trouble. What can you do?

In the United States there is a law that was made to help businessmen in trouble. The law says you can go to a special office in a government building and discuss your problems. A person in the office finds out how much you owe and how much money you have. Together you figure out how to sell the tables and chairs and other things, and then how to divide the income from them among all the people to whom you owe money. Even if you can't pay the whole amount, the law says everybody must accept his share and not try to make you pay more. When you and the person in the government office have settled everything, the law says you are bankrupt.

Bankruptcy laws were first made to help unlucky people, especially merchants. In the days when sailing ships carried valuable cargoes, a merchant could easily lose everything if his ships were wrecked or burned or captured by pirates. Now many countries have special laws that help people in debt. This was not always so. At one time, bankrupt businessmen had to wear special clothes, so that everybody would know they owed money they could not pay. A man could even be put in jail if he couldn't pay his bills.

WHAT IS A PROFIT?

If you get paid for any jobs you do around the house, you get wages. Multiply the number of hours you work each week by the amount you get paid per hour, and you will find out what your weekly wage is.

People get wages or salaries for work they do. But if you buy a skateboard from a friend for $5, then sell it to someone else for $10, you make a profit. How much profit? You can figure it out this way: suppose you usually get paid $1 an hour for the chores you do. And suppose you spend an hour finding someone to buy the skateboard. The skateboard really cost you the $5 that you paid for it, plus the $1 pay for the hour's work you did finding a buyer — that makes $6. Now subtract the $6 which the skateboard really cost you from the $10 you got for it, and the difference is $4. That is your profit. You got this without working for it.

If you put money in a savings bank, you get interest. This interest is also profit. It comes to you, even though you do nothing. Profit comes from owning something — a bank account or a skateboard that you can sell. Profit does not come from working in a bank or having a job in a factory making skateboards.

HOW DOES A BANK MAKE MONEY BY GIVING YOU MONEY?

Suppose you put $10 in a certain savings bank and leave it there for a year. At the end of the year this bank will give you back $10.50. That sounds like getting something for nothing, which almost never happens. How does the bank do it?

The 50¢ extra that the bank gives you is called interest. For every 100 cents you deposit, the bank gives you back 5 extra cents. People at the bank say they pay you 5 per cent interest yearly. Usually this is written 5%.

Now suppose you have a neighbor who needs to borrow $10. The bank will lend her your $10 if she pays the bank 10% interest. That means she must give the bank back $10 plus $1 for interest at the end of the year.

The bank has collected 10% interest from your neighbor. Then the bank gives you 5% interest. You now have 50¢ more than before, and meantime the bank has used your $10 to make an additional 50¢ for itself.

Perhaps you think that is not a big deal. But you aren't the only one who puts money in the bank. Suppose people deposit $1,000,000 in your bank (most banks have much more than that). And suppose the bank lends all of it at 10% interest. The bank will make $50,000 for itself. Of course, the men and women who work at the bank must be paid, and there are other expenses. But even after expenses are subtracted, banks make a great deal of money from savings accounts.

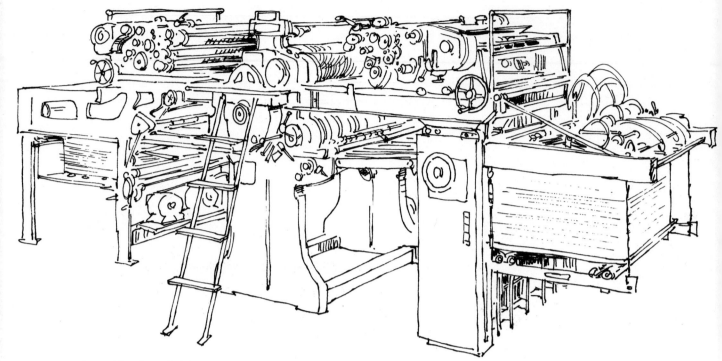

A machine for printing paper money.

CAN MAKING MONEY COST TOO MUCH?

In the United States, the government owns a factory where paper money is printed. Big sheets of very special paper are printed with special ink. Then the sheets are cut up into smaller pieces to make paper bills. Skilled people operate the machines that do the work. Others inspect the sheets. Still others count the sheets and then count the printed bills, over and over, to make sure none get lost or stolen. All of this money-manufacturing is very expensive.

People in the United States use more one-dollar bills than any other kind. It costs the government just as much to print a one-dollar bill as to print a two-dollar bill. So, beginning in 1976, the government produced several million two-dollar bills but fewer one-dollar bills than before. The new money used up less paper, less ink, less working time. And in two years it saved the government about $7,000,000.

WHAT DOES A MILLION DOLLARS LOOK LIKE?

If you piled a million dollar bills, one on top of the other, the pile would be over 625 feet high, higher than a sixty-story building. Placed end to end, a million dollar bills would extend more than 93 miles.

What would a million dollars worth of pennies look like? Piled one on top of the other, the pennies would reach over 63 miles into the sky. They would weigh more than 31 tons.

Yet one small piece of paper can be worth a million dollars. Sometimes people pay for things without using paper money or coins. Instead, they write checks. A check is a note which tells the bank to take money from a person's account and give it to another person or to a company. This note is usually written on special paper. The name of the bank is printed on it and also the writer's name.

If you wanted to buy something for a million dollars, and if you had that much money in the bank, you could write a check to pay for the purchase. But usually only governments and large companies write such a large check. Even so, very few people have ever seen one.

Chapter III

The Mets, the Jets and the Nets

1. What was the four-minute mile?
2. What is a metric mile?
3. What is a batting average?
4. Why is a track field shaped the way it is?
5. Why is a baseball field called a diamond?
6. What is a Quarter Horse?
7. Why are professional football players paid so much?
8. What does love in tennis mean?
9. Which swimming stroke is the fastest?

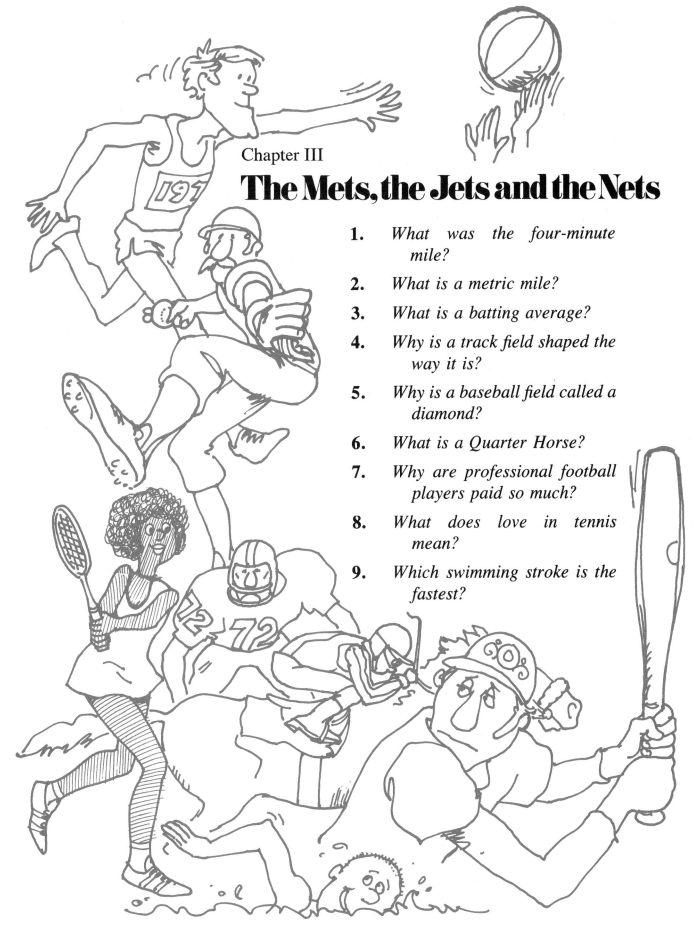

WHAT WAS THE FOUR-MINUTE MILE?

"Roger Bannister breaks the four-minute barrier!"

On July 23, 1954, newspapers all over the world told the story of Roger Bannister's amazing one-mile run. On that day Bannister became the first person ever to run a mile in under four minutes. His record-breaking time — 3 minutes, 59.6 seconds.

Nine years earlier, in 1945, a Swedish runner named Gunder Haegg had run a mile in 4 minutes, 1.4 seconds. Haegg's record had been unbroken for so long that people began to think it might be impossible to run a mile in less than four minutes. Newspaper reporters sometimes called it "the four-minute barrier." Then Roger Bannister ran his historic race.

Ever since Bannister broke the four-minute barrier, the record time for the one-mile race has been steadily falling. What about a three-minute mile? It is hard to believe that anyone could run that fast. Still, many people think that some day runners may break the three-minute barrier. Runners can go faster now because they have studied how the human body works. They train themselves to use their energy in the most efficient way.

The world's record for the mile in 1975, 21 years after Bannister's run, was 3 minutes, 49.4 seconds. John Walker, a New Zealander, set this record. His time was more than 10 seconds better than Bannister's record, and at the rate of 10 seconds every 21 years, someone should run a three-minute mile by the year 2080. Maybe one of your great-great grandchildren will do it!

WHAT IS A METRIC MILE?

In the United States and in other English-speaking countries the most important long-distance race is the one-mile run. But this race is not held in countries where distances are measured in meters. Instead, runners have a 1,500 meter race. Since 1,500 meters almost equals a mile, this distance is called the metric mile.

A metric mile is about 359 feet shorter than a regular mile. So racers run a metric mile a little more quickly than a mile. The world's record for the metric mile, 3 minutes, 32.2 seconds, was set by Filbert Bayi of Tanzania in 1974. The record in the mile race was about 17 seconds longer — 3 minutes, 49.4 seconds. John Walker of New Zealand set that record.

When the United States and England switch to the metric system, metric distances will become official for sports events. But perhaps there will always be a special old-fashioned one-mile race.

WHAT IS A BATTING AVERAGE?

A batting average tells you how often a baseball player gets a hit. You can figure out a player's average by dividing the number of times he has been at bat into the number of hits he has made. A player who is up 9 times and gets 3 hits has an average of 3 divided by 9. But 3 divided by 9 doesn't come out even. It really comes out .3333333 . . . The dots mean that if you kept dividing, you would only get more 3's. Where do you stop? That's the tricky part to figuring out a batting average. You must do the division to 4 decimal places, and then round off to three places. That means you drop the fourth digit if it is less than five. If it is five or more, you increase the third digit by one. The batter who gets 3 hits for 9 times at bat has a .333 average. A batter who has been up 28 times and gets 11 hits has an average of .3928, which rounds off to .393.

The 4th place of a batting average doesn't seem very important, but in 1949 it meant a great

deal to Boston slugger Ted Williams and Detroit third baseman George Kell. That year they tied for the batting championship in the American League. Both hit .343. To break the tie, both averages were calculated to 4 decimal places. Williams hit .3427, but Kell hit .3429 — that is, .0002 higher. So Kell was declared the winner.

WHY IS A TRACK FIELD SHAPED THE WAY IT IS?

A track course has a very special shape. The ends are rounded, but the sides are perfectly straight. Track courses are shaped in this special way because they are used for different kinds of races.

The straight part of the track is for short races, or sprints, such as the 100-meter dash. Sprinters can run a little faster if they don't have to go around a bend while they run.

The curves are for longer races. A stadium with a straight one-mile track could be built, but it wouldn't be practical. People who watch track meets couldn't see the whole race in a stadium a mile long. It would also be possible to run a mile race going back and forth on a short track. But runners would have to come to a complete stop at the end of each lap, the way swimmers do. This would slow the race down. By running on an oval track, runners can keep going for the entire mile. Since the curve of the rounded part of the track is gradual, runners can change direction without slowing down too much.

WHY IS A BASEBALL FIELD CALLED A DIAMOND?

A baseball field is really a square. But when a batter sees it from home plate, the field looks like a diamond. And so that's just what players and fans often call it: "The diamond."

The field didn't have this name when baseball was played a century and a half ago. At that time each town had its own set of rules. In many places, the batter didn't even stand at home plate. Instead, he stood halfway between home and first base. To ballplayers of those days, the field looked more like a square than a diamond.

Around 1845, the players at the Knickerbocker Base Ball Club of New York decided to write down rules for baseball. One new rule placed the batter at home plate — where he stands in today's game. The Knickerbockers' rules caught on. And it wasn't long before people started calling the field "the diamond."

The Knickerbockers made another new rule — one that completely changed baseball. That new rule said that fielders could not throw the ball at base runners. Under the old rules, you *could* throw the ball at a runner, and if you hit him off base, he was out. This meant that baseballs had to be very soft. Usually they were balls of yarn covered with leather.

The new rule made it possible to have heavier, harder baseballs. The new balls traveled much farther when hit. And batters discovered that a well-hit ball could carry over the heads of everyone on the other team. Soon baseball had become the game we know today.

WHAT IS A QUARTER HORSE?

A Quarter Horse is not one-fourth of a horse. Quarter Horse is the name of a special breed of riding horse. Thoroughbred is another breed.

The Quarter Horse gets its name because Quarter Horses are the fastest horses in the quarter-mile race. The breed was first developed in the United States over 200 years ago. In those days race tracks were rarely longer than a quarter mile because they usually had to be cut from dense forests. So Quarter Horses were bred to run the quarter-mile distance faster than any other horse.

Today cattlemen in the American West use Quarter Horses for work that requires quick starts and fast turns.

WHY ARE PROFESSIONAL FOOTBALL PLAYERS PAID SO MUCH?

Amateur athletes don't get paid. They are in sports for fun or just because they want their teams to win. But a professional athlete earns money — sometimes a great deal of money. Professional athletes are really entertainers — like movie stars or singers. People love to be entertained, and so in many parts of the world they reward those who give them pleasure.

The best football players are among those who get the most money. They know they have valuable skills, and they put a high price on their work. In the United States the biggest stars make hundreds of thousands of dollars a year. Many singers and movie stars do, too. These actors and musicians can keep on working till they are old, but athletes lose their strength and skill as they get older. Then they have to quit their jobs. So they often don't earn as much in a lifetime of work in sports as some people earn in other professions.

WHAT DOES LOVE IN TENNIS MEAN?

The game of tennis was invented so long ago that no one has been able to find out why the score is kept the way it is. When a player makes his first score, he doesn't get one point — he gets 15. His next score gives him 30. His next gives him 40.

Suppose you have made your first score and your opponent has no points at all. Do you say, "It is 15 to nothing in my favor"? No. You say, "Fifteen love." In tennis, "love" means "zero." But why?

Most people think that love was a slang word to begin with. The French players who invented the game probably said "Fifteen for me and an egg for you" when they meant that the score was 15 to 0. In those days, "egg" was a slang word for "zero." The French word for egg is "l'oeuf." And so when Englishmen learned to play the game they probably said something like "luff" at first, and then changed it to love.

Everybody wanted to play tennis when the game first came to England about 700 years ago. Lords, priests, soldiers, and even some common people spent a great deal of time hitting the little cotton ball back and forth. Then, in 1388, a law was passed forbidding tennis games. The king was afraid he would lose the next war if soldiers played too much tennis. He thought they should spend all their time practicing archery, since wars were fought with bows and arrows in those days. So tennis became illegal. Later, the law forbidding tennis was changed. But when people began to play again, they kept on using the word *love*.

WHICH SWIMMING STROKE IS THE FASTEST?

Around the year 1900 almost everyone who could swim simply paddled along with hands and feet. This "dog-paddle" stroke was slow. A few good swimmers used the breast stroke, a way of swimming which had been known for centuries.

Then, in 1902, two brothers from Australia named Charles and Syd Cavill arrived in England with a new way of swimming. The stroke they brought with them probably originated in the Pacific islands to the northeast of Australia. When the Cavills swam, they looked as if they were crawling through the water. So their stroke was called the Australian crawl. It was much faster than any other stroke, and it quickly became popular in England.

The next year the brothers brought the crawl to the United States, where it was also an instant success.

Today the Australian crawl is still the fastest stroke. Whenever there is a free-style race — a race in which any stroke is allowed — almost everyone swims the crawl.

Chapter IV
Fortunetelling and Looking Into the Future

1. *Were you born under a lucky star?*
2. *How did astrology begin?*
3. *Is a comet an evil omen?*
4. *What do people mean when they say "I've got your number"?*
5. *How magical are magic squares?*
6. *Can a mind reader really read your mind?*
7. *What is ESP?*
8. *Can cards tell your future?*
9. *Are there any real ways to predict the future?*
10. *What is the census?*
11. *How do public opinion polls work?*

WERE YOU BORN UNDER A LUCKY STAR?

Have you ever wondered what your future will be? People have always looked for ways to know the future. Sometimes they turned to the stars for answers to questions that really had nothing to do with the stars.

Making predictions by the stars began thousands of years ago in many parts of the world. In ancient Babylonia and Egypt priests watched the skies all through the night. From their observations they learned that many stars disappear for part of the year, then reappear. One after another, they rise in the east, each at a particular point on the horizon. Each year, each one shows up at the same date. Because of this, the stars could be used to foretell the coming of winter, of spring floods and other annual events.

In Egypt the priests recognized the star that rose right before the yearly flooding of the River Nile. When this star was sighted, the priests would tell farmers to open irrigation canals so that the floods could be used to water their fields. Other stars were used to tell farmers when to plant seeds and when to harvest. Life depended on knowing these dates. So people thought the stars that announced them brought good fortune to everyone.

The idea that a star could bring luck to an individual came later. Kings, for example, began to demand personal forecasts. So priests made them, and based their predictions on the stars that rose on royal birthdays. Eventually, lots of other people began to have their fortunes told in the same way.

In modern times, scientists have learned what the stars can and cannot predict. The stars can be used as a calendar, and they can warn us of changes in seasons. But there is no proof that the stars have any connection with the daily lives of individuals. The star you were born under is neither lucky nor unlucky.

HOW DID ASTROLOGY BEGIN?

The kind of fortunetelling now called astrology began in ancient times. It grew out of the knowledge that stars could be used to foretell yearly floods and changes in seasons. From this idea people went on to believe that stars could foretell other events and influence their lives.

More than 2,500 years ago the Greeks thought that twelve groups of stars, called constellations, were especially important. The names of these constellations came from old legends: Taurus the Bull, Gemini the Twins, Cancer the Crab, Leo the Lion, Virgo the Virgin, Libra the Scales, Scorpio the Scorpion, Sagittarius the Archer, Capricorn the Goat, Aquarius the Water Carrier, Pieces the Fish, and Aries the Ram.

The sun, moon, and planets were also considered important. In fact, the Greeks gave them the names of gods. Some people believed they *were* gods. All these heavenly bodies together were supposed to control the lives of everyone.

Sky watchers noticed that the twelve constellations and the sun, moon, and planets always appeared in a certain zone of the sky. They called this zone the *zodiac*. That was because zodiac meant *circle of animals,* and over half of the constellations had animal names. In keeping track of the heavenly bodies, the sky watchers divided the zodiac into twelve equal parts called houses or signs. Each constellation was located in a house. A diagram of the zodiac — a sort of map showing the houses — was drawn like a pie cut into twelve slices.

The sky watchers learned that the sun's daily

An ancient Egyptian record of the star Sirus which appeared before the River Nile flooded farmers' fields.

path shifts north for part of the year, then south, then north again. On any given day the sun will rise in the part of the sky where one of the zodiac constellations is at dawn. In spring the sun is "in Taurus." After about thirty days, its path has shifted, and the sun is "in Gemini." It appears in each zodiac constellation for about a month. At the end of the year, it is back to Taurus again. In ancient times, many people believed that the sun god was actually visiting the constellations.

Greek fortunetellers thought the heavenly bodies influenced a baby at the moment it was born. The child's whole life would depend on which house the sun was supposed to be visiting that day. For example, if the sun came up "in Taurus," the child would be stubborn, like a bull. The positions of the moon and planets at the moment of birth would also affect the child's future. By looking at the twelve-part sky map, it was possible to see where the heavenly bodies were on a baby's birthday. Then the fortuneteller would predict its future, and these predictions were called a horoscope.

Fortunetellers could imagine any story they liked about star-gods, for no one knew much about the heavenly bodies. The universe seemed mysterious and scary to most people. They often felt very helpless. They wanted some way of knowing what would happen to them. Horoscopes seemed to be the answer to their need.

The sky watching that the Greeks did became known as astrology. Its main purpose was fortunetelling. But after a while people began to make a scientific study of the heavens. They tried to find out what the stars really are and why the sun moves about in the sky. These people became known as astronomers.

But belief in astrology continued for many centuries. Horoscope-making spread to many countries. Meanwhile, the records of astronomers became very accurate. By 1492, the year Columbus sailed to America, astronomers knew that when the sun rises and sets, it only *seems* to travel around the earth and across the sky. They had discovered the truth — the earth turns on its axis and makes a yearly trip around the sun.

Other scientific discoveries followed quickly. Astronomers learned that the sun is 93 million miles away from the earth, and that the stars are trillions and trillions of miles away. The sun could never visit the constellations!

Astronomy has become a very important science. Airplane pilots depend on it to get safely around the world. Astronomers helped to make it possible for men to land on the moon and even to bring back samples of moon rock. Of course, astronomers realize that it would be comforting to depend on a horoscope. But they can find absolutely no way that fortunes can be told by the stars. What do you think?

"Comet pills — half price — going out of business"

IS A COMET AN EVIL OMEN?

A strange thing sometimes appears in the sky. A fuzzy patch of light grows larger night after night and develops a bright tail. This is a comet. It seems to be rushing toward the earth.

No wonder people were once afraid of comets. In the days when stars were used to predict events, unpredictable comets were thought to be evil omens. People believed they caused plagues, famines, and even wars.

These ideas changed when scientists began to understand gravity, which explains how heavenly bodies move. Isaac Newton had just discovered the law of gravity when a brilliant comet appeared in the sky in 1682. Many people were frightened by it. But the comet was a challenge to Edmund Halley, an English astronomer. He observed it carefully. Using Newton's Law, he worked out its path and predicted its return. Halley said the comet would reappear in 1758. It *did* return, and ever since then it has been called Halley's comet.

When the comet reappeared in 1910, peddlers of patent medicine sold concoctions they called "comet pills." The pills, they claimed, would protect people from the comet's dangerous and evil influence.

There will be little talk of evil omens when Halley's comet appears again. Millions of people will enjoy watching it grow brighter in the sky. Probably you will be one of the watchers, because Halley's comet is due to return in 1986.

WHAT DO PEOPLE MEAN WHEN THEY SAY "I'VE GOT YOUR NUMBER"?

Long ago people believed a special number went with each person's name. This idea probably started over two thousand years ago when the Hebrews and Greeks used the letters of the alphabet for numbers. Each letter was given a number value. The value for A was 1, for B it was 2, and so on. Finding a person's number was easy enough. It just meant adding up the values of the letters in his or her name. If the sum was believed to be a lucky number, the name was supposed to bring good fortune. Ada, for instance, gave a total of six — considered a very lucky number. A big sum was also a sign of good fortune — and the bigger it was, the better. A name like Solomon, which had a high value, was supposed to give a child an excellent start in life.

Yet a good name did not guarantee success. Someone might put a curse on the name by misspelling it and thereby change its total value to an unlucky number. Whoever did that could say, "I've got your number," which meant, "I am in control of you." The expression still means that, more or less. That's why teachers sometimes use it. It's a way of saying, "I'm in charge here and I know what you are up to. You can't fool me."

HOW MAGICAL ARE MAGIC SQUARES?

One day, over 4,000 years ago, the Emperor Yu of China was sailing on the Yellow River. Not far from the ship was an unusual sight — a turtle with Chinese numbers painted on its shell. There were nine numbers, arranged three in a row. What did they mean? Were the numbers an omen? The Emperor tried adding them. He added them across, up and down, and diagonally. Each time the sum was the same — 15. The numbers seemed to have a strange magic power. Surely they would ward off evil. The Emperor ordered a charm made with the numbers on it, set in a square. He believed the charm really worked, and so he wore it all the time. That's how magic squares began — or so the story goes.

At least one part of the story is true, for magic squares did start in China. From there they spread all over the world. In India they were hidden at the bottom of bowls that were used in fortunetelling. Muslim doctors drew them on the feet of their patients to ward off the plague. Eastern Jews made them into religious symbols by cutting off the corners. Their sum was 15, which was a very sacred number because it stood for the name of the lord Jehovah.

Some squares were used by people who believed that certain numbers stood for the sun, moon, and planets. The number for Mars, for instance, was 9, and Mars was the protector of maidens. The number 9 appeared in a square that added up to 18. Since 1 + 8 = 9, this square was put on a charm to be worn by young girls. This charm was very popular in the Middle Ages.

Magic squares still turn up in many places. They are used as problems in puzzle books. In Africa women wear them on beautiful silk scarves. On some cruise ships they are painted on the decks so that tourists can play shuffleboard. The numbers on the shuffleboard happen to be the ones that the Chinese used in the first magic squares.

CAN A MIND READER REALLY READ YOUR MIND?

If anyone tells you he can read your mind, watch out. He is about to play a trick on you. "Pick any number under fifty," he might say. "And to make it a little harder, pick a number with two different digits, both odd—31, for example. You're doing it! I can tell from your thought waves. I can tell that you're not picking eleven because it's made of two ones, and both digits are the same. And you're not picking two because two is even. Okay, write down the number you've picked and concentrate on it. I've got it — it's 37! Right?"

And very likely that's the number. Why? Well, how many numbers under 50 are made of two different odd digits? Only eight: 13, 15, 17, 19, 31, 35, 37, 39. There is only one chance out of eight that the mind reader will guess right. But he improves his chances by steering you away from the numbers beginning with one. That leaves only four choices and reduces the performer's chances to one out of four. Then he mentions 31. That steers you away from 31. With only three choices left, you and almost everybody else pick 37. Seven seems to be most people's favorite number. So nearly nine times out of ten, the number chosen is 37.

Magicians call this type of trick a *force*. It is used by almost every mind reader.

Another common mind reading trick depends on a switch. This is often done with money. "Does anyone have a dollar bill I can borrow?" the mind reader asks. "Please hold it up." You offer him one. The performer takes it and tells you how to fold it so the serial number won't show. While checking the fold, he gives you a pad and tells you to write down the number of the bill. "Concentrate on it," he says. "Now, from your thought waves I can tell that the number is. . . ." And the mind reader is right! Why? Because it's the number on a dollar bill of his own. Before the show, he memorized the number, folded up his bill, and put it in his pocket. As he talked to you, he used sleight of hand to switch his bill and yours.

For a longer trick, the mind reader may ask people to show his assistant some objects such as a knife, a watch, or a ring. Then, without ever seeing them, he names them correctly, one after another. How? His assistant signals to him in some sort of code.

All three methods — the force, the switch, and the code — were among those used by Joseph Dunninger, a famous mind reader. Dunninger claimed he could tell what others were thinking of, and he was such a good showman that audiences believed him. When he performed, he called himself a "Master of Mental Mysteries." But those in show business knew he was just a very good magician.

WHAT IS ESP?

All the mind reading acts in shows are fakes, but does that prove that mind reading is impossible? Does this ability exist? People who are called parapsychologists are doing experiments, hoping to prove the answer is yes. They believe there is a way of getting information without using the senses of sight, hearing, smell, taste, or touch. They have given this a long name — *extrasensory perception*. They call it ESP for short.

One type of ESP is called precognition. This is supposed to be a way of finding out about events before they happen. The people who make sensational predictions about the future call themselves psychics, and they say they have precognition. Many of their predictions are clever guesses that anyone could make. When they come true, these are remembered. The others are forgotten.

Ever since the study of ESP began, fakers have told stories and given demonstrations that seemed convincing. About 100 years ago many scientific investigators believed that two young Englishmen, Douglas Blackburn and G. A. Smith, could transfer thoughts from one to the other. In one famous demonstration, a scientist scribbled a drawing and gave it to Blackburn. Smith was sitting on a chair, completely covered with thick blankets, his ears plugged. After ten minutes he reached out a hand for a pencil and paper. Then he suddenly threw off the blankets and jumped up, waving a copy of the scientist's drawing. It was a good show, but it was not a proof of ESP.

Thirty years later Blackburn explained how he and Smith did the trick. Blackburn had somehow managed to copy the drawing and to stuff it into a mechanical pencil which was given to Smith when he stretched out his hand. Since Smith was buried under blankets, no one could see that he used the glow of a luminous slate to copy the drawing he found inside the pencil. That's all there was to it.

Other demonstrations turned out to be hoaxes, too. No method of studying ESP seemed to be any good until Joseph Rhine in the United States started testing people with cards. He used a pack of 25 cards marked with five different symbols: a circle, a square, a plus sign, wavy lines and a star. The idea behind these tests was simple enough. Anyone who tried to name the cards as they were dealt from the pack had one out of five chances of being right. He probably would guess 5 out of 25 correctly. So Rhine reasoned that anyone who averaged better than that on several trials must be getting information by ESP. When the tests were given to hundreds of people, Rhine did find many who averaged more than 5 out of 25 correct answers. He announced this in 1934.

Others repeated Rhine's tests and some of them got similar results. Several rich people became interested and gave money for research on ESP. Those who took the tests got so many good scores that some scientists became suspicious. A careful check of the people who had scored very high showed that some of them were fakers. In one way or another they had found out what cards were turned up, and then they had pretended to get their information by ESP.

In 1974 one of Rhine's own staff members was caught fixing up the figures in an experiment. So far, no one has been able to devise foolproof experiments. This makes many scientists doubt there really is such a thing as ESP.

"I foresee trouble with fractions."

CAN CARDS TELL YOUR FUTURE?

Madame Cartalinka, the fortuneteller, is inviting people into her booth at the fair. You decide to find out what she has to say.

As you sit down, Madame Cartalinka hands you an ordinary deck of 52 playing cards. "Your future is in those cards," she says. "Shuffle them, then deal out twelve cards. Put them in four rows, three in a row." After you turn up the cards, she studies them. "Mmmm — two kings side by side," she says. "That means you will be talking to officers of the law. But don't worry. You're not going to get into trouble. The officers probably will be men who check passports and things like that." After a pause she says, "See those two eights? They mean you'll take a trip. That queen of hearts over there is a sign you will meet a fair lady in your travels. . . ." And on she goes, explaining each card.

By the time your fortune is finished you recognize Madame Cartalinka. She is your neighbor, Eleanor Jones. She admits she is telling fortunes for fun, but she says she uses a system that is centuries old.

There are several age-old systems for reading cards, but they are all alike in one way. The meaning of each card depends on its position, its suit and its value. If 12 cards are used out of a deck of 52, the number of different combinations possible is really enormous. It is about 20,000,000,000,000,000 — 20 quadrillion! A person has very little chance of turning up the same 12 cards again in the same order. So there's practically no chance of getting the same fortune a second time. In fact, the second reading usually contradicts the first. The cards themselves have no value in predicting the future. And if a fortuneteller tells you things that come true, it is because she has cleverly put together events that might happen to anyone. If some of them happen to you, you will probably forget the ones that don't work out.

ARE THERE ANY REAL WAYS TO PREDICT THE FUTURE?

Have you ever seen an eclipse? Many years ago people were frightened by eclipses. They thought the gods were blowing out the sun. No one understood why the sun sometimes disappeared, then began to shine again. And no one could tell when this mysterious event was going to happen.

Today astronomers can predict every eclipse. The study of the stars, planets, and other objects in space has led to the discovery of mathematical laws that describe their movements. Mathematics is the tool that astronomers use to make accurate predictions.

Physics is another science that depends on mathematics. Physicists, like astronomers, can predict the future. They can tell us how fast things will fall or how hot they will be when they burn.

Most scientists try to find laws that fit our universe, our world, and ourselves. When they can use mathematics, their predictions are accurate. Astronomers and physicists were the first scientists to discover this. Other scientists are trying to catch up.

Psychologists study the human mind. They have been looking for mathematical laws to predict how human beings will think and behave. So far they have not had much success, because people's minds are extremely complicated.

Would you like to be able to predict the future? One sure way is to study astronomy or physics.

WHAT IS THE CENSUS?

In recent times there was a rich man who owned some houses that he rented to other people. He decided he could make more money if he had even more houses, but what kind should he build? Should they be small — or big enough for families with four children? Was there some special kind of building that he could be sure of renting? To find out, he did some detective work. And a bright idea came to him. He discovered that men and women were living longer. Every year there were more old people than before. Why not build special apartment houses where only old people would live? That was what he did.

The businessman got his information from a census report. A census is a record that a country's government makes. It shows how many men, women and children there are, where they were born, what their ages are, what their jobs are, whether they have automobiles, washing machines, bathtubs, television, whether they can read and write. All these facts — and many more — are especially useful for planning ahead. The number of children tell how many new schools will be needed. Figures about health tell whether there should be more hospitals. Bathtub manufacturers want to know how many they can expect to sell.

Taking a census is a big job. Some countries do it every five years, others less often. The United States government must count the people in each state every ten years. If a state's population grows a great deal, then that state will have more representatives in Congress. The United States was the first country to use the census in this way.

The census in other places was often just a list of men who could be made to pay taxes or serve in the army. To avoid paying or fighting, most people tried to escape the census takers who went from door to door writing down names in official books.

Some records are still made by census takers who ask questions and write down answers. That takes time. So, to speed things up, many people in the United States receive printed questions in the mail. They check off the answers and send the forms back. Computers then count the number of children, radios, bathtubs, and other things that are listed. United States law says that names on census forms must be kept secret. Nobody, except you or your nearest relative, can ever look up your name in census records and see what you said on your report.

On the day when the census is taken in the country called Sri Lanka, these women and their families must stay indoors to be counted.

"You like spinach, don't you?"

HOW DO PUBLIC OPINION POLLS WORK?

Suppose you want to find out how many children in your school like spinach. There is one way to get a fairly accurate answer. Give each child a plate of spinach and see what happens.

But this will take a long time if the school has two thousand students. So instead of offering each child some spinach, you decide to ask a few children if they like it. In other words, you will take a poll.

Suppose you find that half the children you poll say they do like spinach. What does that really mean? Well, let's see.

Your first problem in taking a good poll is picking the children to be questioned. If you asked everyone in the eighth grade, they would all be about the same age. But younger children would be left out of your poll. And they are the ones who often don't like vegetables at all.

The best plan is to give children from every class a chance of being picked. You can do this by writing the name of each of the two thousand children on a separate slip of paper and putting all the slips in a barrel. Then you remove fifty names at random. Now you poll the fifty children whose names you have picked. This method of choosing the children is a lot of trouble, but it will give you the most accurate results.

Still, there are other problems to solve when you take a poll. Will the children answer your question honestly? Answers sometimes depend on the person asking the questions. A child may give one answer to another child and a different answer to an adult. A boy might not give the same answer to a man as to a woman.

Experienced pollsters know that answers often depend on the way a question is asked. Suppose a pollster says: "You like spinach, don't you?" A child may say, "Yes" without thinking. But "Do you like spinach?" might bring a more honest answer.

And there are probably some children who don't care one way or the other about spinach. For them, "Yes" and "No" are both wrong answers. But they might be more likely to answer "Yes" than "No." So pollsters often ask people to answer "Yes," "No," or "I don't care."

You can see that taking a good poll is not easy. There are many reasons why polls sometimes give the wrong results. For example, they do not always answer correctly, "Who is going to win the election?" But if they are done carefully, polls are fairly accurate most of the time. And they can tell a great deal about the way people think and feel.

Chapter V

Numbers: The Great Invention

1. *Why do we call numbers a great invention?*
2. *How did names for numbers get started?*
3. *Where did our numerals come from?*
4. *How did the Romans add with Roman numerals?*
5. *Which came first: zero or one?*
6. *Who were the zero freaks?*
7. *Can you count without numbers?*
8. *Is zero the same as nothing?*
9. *Can anything be less than nothing?*
10. *What is the smallest number in the world?*
11. *Did people ever count on their toes?*
12. *Do you remember how you learned to count?*
13. *Why do astronauts use "count-down" instead of "count-up"?*
14. *Can any animals count?*
15. *Why do we count eggs by the dozen?*
16. *Who invented fractions?*
17. *What are decimals?*
18. *Who invented decimals?*
19. *What is a round number?*
20. *What is a square number?*
21. *Are there any triangular numbers?*
22. *Does 1 + 1 always equal 2?*
23. *Do all people count the same way?*

WHY DO WE CALL NUMBERS A GREAT INVENTION?

One of our greatest inventions has been put together without any nuts or bolts or gears. It burns no fuel and makes no noise. It is easy to carry around, although it is made of so many parts that no one can count them all. These parts never need repairs. They never wear out, no matter how much they are used.

Every nation in the world now makes use of this invention. Yet not a single person has ever seen it, or touched it, for it is not a machine. It is a system that is made of ideas, not pieces of metal. Its parts are numbers — the ones we use in counting, measuring and weighing.

At first, it may seem strange to call our number system an invention. But, like the automobile and television, this system was created by people. It did not happen all by itself. And, like many other inventions, it has been improved from time to time. Some kinds of number that we use today — such as fractions — are not nearly as old as the ones we count with. New ways of using numbers have been invented, too.

Today there are so many different uses for our number system that almost everything we do depends on it. Our number system has been used in planning and making nearly all our other inventions. No wonder it is called one of the greatest inventions in the world!

HOW DID NAMES FOR NUMBERS GET STARTED?

Using names for numbers began so long ago that no one knows exactly how it started. But math detectives have found a few clues that tell what may have happened. In the beginning people had no names for numbers. Instead, they probably counted by bending first one finger, then another. Eventually, the name for the first finger became the word for one. The second finger name became the word for two, and so on.

Some American Indians long ago named numbers in a similar way. They said, "The end is bent" for the number one, and by that they meant the little finger was bent. "It is bent once

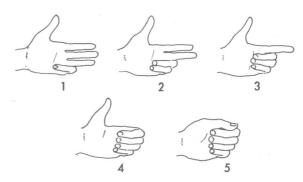

more" stood for two. "The middle is bent" meant three. "Only one remains" was four. "My hand is finished" meant five.

In English our fingers are called digits, and so are the numbers zero through nine. The Persian word for five is *pantcha,* which is very close to the Persian word for hand — *pentcha.*

And in Russian *pyat* means five and *pyad* means "hand with five fingers spread out." These clues are among the few that tell us how number names got started.

Other very early names have been changed so much that it is very hard to tell what they were thousands and thousands of years ago.

WHERE DID OUR NUMERALS COME FROM?

Long before people could write words, they found ways to keep track of things. They made scratches or dots or notches on whatever was handy. A hunter would take a bone and make a mark for each deer he had killed. Or he might make a mark on a stick for each day that had passed since the new moon appeared. One dot or scratch meant 1; two dots or scratches meant 2; and so on. These tally marks probably were the first numerals.

After the Hebrews and Greeks got an alphabet, they started using letters for numerals. The first nine letters stood for the first nine numbers. The next nine letters stood for 10, 20, and so on, through 90. Special signs and combinations of letters were used for higher numbers.

The Romans invented a better system. They may have developed it from tallies. One scratch stood for 1, two for 2, three for 3, four for 4. Crossed scratches —X— stood for 10. Since half of 10 is 5, the top half of X — that is, V — stood for 5. Scratches in other shapes stood for 50, 100, and 1,000.

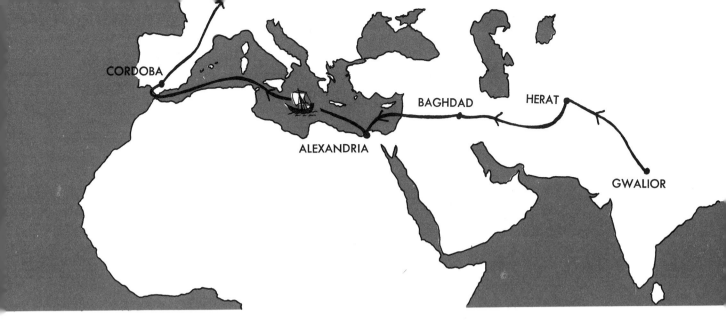

The route along which Hindu numerals spread from India to Europe.

About 2,000 years ago, Roman armies fought and won battles all over Europe. After conquering lands, they set up governments. This meant keeping records, and Roman numerals were used in keeping them. In this way the Romans introduced their numerals to many other people.

Meantime, in far-off India, some Hindu people worked out a way to write numbers with ten special signs. One of these signs was zero. Zero was a wonderful invention. Using zero, people could solve much harder problems than they had ever solved before.

You might think this Hindu system would have spread quickly to other parts of the world. It didn't. Hundreds of years went by before people in Europe learned about it.

Hindu numerals reached Europe in a roundabout way. Over 1,300 years ago, an Arab named Mohammed was born. Mohammed became the leader of the Islamic (or Muslim) religion. At first, all his followers lived in Arabia and spoke the Arabic language. But then Muslim armies began to conquer other people. They went as far as India in Asia and all the way across Africa to Spain in Europe.

Wherever they went, they picked up new ideas and inventions. When the Muslims reached India, they learned about zero and the nine other Hindu number signs. They started using them and began to write about Hindu arithmetic. Of course, they wrote in their own language, Arabic, but they shaped all the numerals as the Hindus did.

At that time very few people in Europe could speak or read Arabic. Those who saw the remarkable Hindu-Arab arithmetic books ignored them. But about 800 years ago one book was translated by an English monk. And in the year 1148 a hand-written book appeared in Germany giving the Hindu multiplication table up to 9×9. At last, the nine signs and zero began to spread across Europe. But, by this time, everyone had forgotten that the Muslims had borrowed the Hindu's numerals. People who used them called them Arabic numerals. That's what we call them today, although we often leave out the word Arabic and just call them *our* numerals.

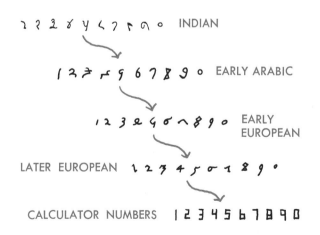

How the Hindu numerals changed as they spread to Europe.

HOW DID THE ROMANS ADD WITH ROMAN NUMERALS?

They didn't. The Romans used their numerals only for keeping records. They could build fine temples and great roads, but they never managed to work out a way to add with their numerals.

Writing them was hard enough. For example, take the numeral eighteen. That was formed like this: XVIII. The X stood for ten; V stood for five, and I meant one. Writing twelve was a little easier — that was just XII.

But no Roman would bother to add XVIII and XII. That kind of problem was done by moving pebbles around in grooves on a counting board. If a pebble was placed in the groove at the right, its value was one. If it was in the next groove, its value was five. In the third groove, its value was ten.

A Roman set up eighteen (XVIII) in this way:

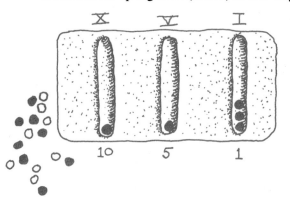

To add twelve (XII), he put another pebble in the X groove and two in the I groove. Then the board looked like this:

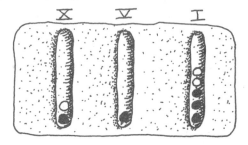

The next step was to see if several pebbles in one groove could be replaced by a single pebble from the next higher groove. That meant taking five pebbles from groove I and putting one pebble in groove V. Now the counting board looked like this:

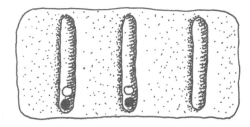

But two fives equal ten. So the Roman took two pebbles out of the groove. He replaced them with one pebble that he put in groove X. Finally, he had the answer — XXX, or, as we would write it, 30.

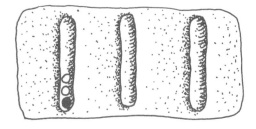

No rich Roman ever touched the counting board. A slave worked at it, doing arithmetic for his master.

People stopped using Roman counting boards many centuries ago. Arabic numerals made them unnecessary. But, in a way, we still refer to the boards when we do arithmetic. Our word *calculate* comes from *calculi,* and that was the Roman word for pebbles.

WHICH CAME FIRST: ZERO OR ONE?

About four thousand years ago, Babylonian children went to school and learned to use numbers. Instead of writing on paper, they made marks on flat pieces of wet clay. Their number system had the number one, but it did not have zero. No Babylonian child ever got a zero in school. It took a long time for people to realize that the question "What is one minus one?" could have an answer: $1 - 1 = 0$.

The idea of zero grew from the idea of emptiness. More than a thousand years ago the Hindus used the word *sunya* in their numbers. *Sunya* meant empty space. It stood for an empty place in a number, such as 403. But at first no one thought of it as a number. After many years people realized that *sunya,* or the empty space, was also a number, which we call zero.

The number one was several thousand years old before the number zero was invented.

WHO WERE THE ZERO FREAKS?

Everywhere in Europe people now do arithmetic with the Arabic numerals: 0, 1, 2, 3, 4, 5, 6, 7, 8, 9. This wasn't always so.

Europeans first heard about Arabic numerals nearly 800 years ago. Until then, they had used Roman numerals for keeping records. To do arithmetic they used an abacus or a counting board — or their fingers. Some merchants and traders could add and even multiply very fast on their fingers.

These old ways of doing math were good enough for everyday business. For a while, nobody wanted to change over to the Arabic system. The numerals looked strange. And the idea of zero was just too baffling. By itself, zero stood for no quantity: $0 + 5 = 5$. But if you placed zero after 5, you got 50! Was zero a digit — or some kind of magic? The question caused much argument. And if you wanted to annoy someone, you called him a "zero-freak."

Slowly bookkeepers discovered that they could do their work faster with Arabic than with Roman numerals. Then one sly fellow saw that he could easily cheat with the new figures. He could change a 5 to look like a 9 and make extra money for himself if a customer did not notice. In those days the numerals were shaped a little differently from modern ones. One jog of the pen made $\mathcal{G}$ into $\mathcal{9}$. With Roman numerals it was not so easy to change V to IX.

Soon more bookkeepers in the city of Florence in Italy learned the trick. But they were found out. The city passed a law saying that nobody could use Arabic numerals. Later, of course, the new numerals became lawful again because they made work so much easier.

CAN YOU COUNT WITHOUT NUMBERS?

Workers in a pencil factory can learn how to pick up exactly twelve pencils from a pile. They don't have to count before they put the pencils in the box.

Sometimes only one look is needed and then you can tell how many things are in a group. But that's not the same as counting. You need numbers in order to count.

In counting objects, you start with one and go on from there, matching each thing in the group with a number that is larger than the one before it. You need to know that after two comes three, and after that comes four, and so on. If you make a mistake, you will wind up with the wrong total. You have to put the numbers in the right order or you are not counting.

IS ZERO THE SAME AS NOTHING?

When you say 3 − 3 = 0, you are using zero as a number that stands for "nothing." When you add zero, you add nothing. When you subtract it, you take away nothing. When you multiply by it, or divide into it, you get nothing.

We also use zero in a different way. When we write 307, we are writing a numeral that means 3 hundreds, plus 0 tens, plus 7 ones. In the numeral 307 the digit 0 is a marker that tells us 307 is different from 37. In this case, zero is not the same as "nothing."

CAN ANYTHING BE LESS THAN NOTHING?

In Canada the weather man may say, "Sunny today. Temperature: minus two degrees Celsius." In writing, that would be −2°, which means 2 degrees less than zero. Zero degrees Celsius is the temperature at which water turns into ice.

We call a number like − 2 a negative number. We put the minus sign in front of it to show that it is different from the numbers above zero.

Negative numbers are often used in keeping track of money. Bank accounts sometimes can have less than nothing in them. If you have −$50.00 in your account, you owe the bank $50.00. The bank charges you interest on the money you owe until you pay it back. Meanwhile, the bank is making a profit on the negative amount of money in your account!

WHAT IS THE SMALLEST NUMBER IN THE WORLD?

Do you think it is zero? There are numbers smaller than zero. They are the negative numbers, such as minus one, minus two, minus three. Imagine a straight line that you can extend at both ends as much as you like. Pick a point on the line and call it zero. The positive numbers can be marked off to the right of zero. To the left of zero are the negative numbers. The farther to the right a number is, the larger it will be. As you move to the left past zero, the numbers grow smaller. Any negative number is smaller than any positive number. Minus one is smaller than plus one. And minus fifty is smaller than minus forty-nine.

What is the smallest number you can think of? Whatever it is, there is a number to the left of it that is even smaller. This means there can be no smallest number.

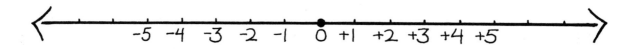

DID PEOPLE EVER COUNT ON THEIR TOES?

Long ago the Tacamanaco Indians in South America counted on their fingers and their toes. They used their fingers to get from one to ten. A Tacamanaco counted, "One, two, three, four, a whole hand." "A whole hand" meant five. "One on the other hand," meant six. "Both hands" meant ten. For eleven the Tacamanaco said, "One on the foot." "Two on the foot" meant twelve, and "a whole foot" meant fifteen. When they had used up all the toes on one foot, they began counting on the other foot. "One on the other foot" meant sixteen. For twenty they said "A person." How did the Tacamanacos count past twenty? Twenty-one was "One on the hand of another person."

DO YOU REMEMBER HOW YOU LEARNED TO COUNT?

Scientists say nobody remembers learning to count. Many little children learn to say the numbers one, two, three, four, and so on, before they really can count. But they are only repeating the sounds of the words without knowing what the words mean. It is the same as saying "hey diddle diddle" or "fee, fie, foe, fum." Perhaps when you were small you could say all the number words in proper order from one to twenty. Yet you may not have known which is bigger — fifteen or eighteen.

It takes time and practice to understand number words. First, a child must be able to compare two things and tell which is bigger, or longer, or heavier. This may sound very easy, but it isn't easy for a baby. Sometimes a baby tries very hard to put a large pot into a smaller one, and then starts crying because the large pot doesn't fit. You did a lot of experimenting before you learned the meaning of smaller and larger.

Next, a child has to learn to put several things in order. He or she does this in different ways. Longest to shortest is one way, and lightest to heaviest is another.

Scientists think that after you learned to put things in order, you were ready to understand numbers. Soon you could tell in what way three spoons were like three plates. And that was when you really began to count.

WHY DO ASTRONAUTS USE "COUNT-DOWN" INSTEAD OF "COUNT-UP"?

The first count-down was in a German movie called "The Girl in the Moon." This was one of the earliest science fiction films. It was shown in 1928, long before any real space rockets had been launched.

In making the movie, the director wanted to dramatize the moment the rocket blasted off from the earth. The usual "1-2-3 — go!" signal seemed too tame. So someone suggested counting down instead of up. To make it even more impressive, the director had the count-down start at 10. At zero, the fake rocket left in a cloud of smoke. This turned out to be the best scene in the movie, and it was widely copied in science fiction films that followed. Space fans began to think it was part of a real launching.

When scientists started experimenting with space rockets, they found the count-down a useful way of checking before the final firing. And now astronauts and everyone else who works with rockets use the count-down.

CAN ANY ANIMALS COUNT?

Some wild animals can tell that a group of objects becomes different when the number of objects in it has changed. Many birds have this number sense. If a bird's nest contains four eggs, and one is removed, the bird will stay at the nest. If two eggs are removed, the bird usually will desert the nest.

Wasps also build nests for their eggs. Some of them always provide caterpillars for the baby wasps to eat when the eggs hatch. The mother wasp supplies each egg with exactly the same number of caterpillars. Some kinds of wasp always provide five caterpillars for each egg. Other kinds always provide twelve or twenty-four. In one species of wasp the male is much smaller than the female. Somehow the mother wasp knows whether a male or female baby is going to hatch from an egg. The mother provides the eggs containing males with five caterpillars, and the eggs containing females with ten.

This number sense in wild animals does not involve putting the numbers in order. So it is not counting.

But do any trained animals count? Suppose a trainer says to a dog, "How much is ten and five?" The dog begins to bark. When he has barked fifteen times, a slight motion of the trainer's hands or eyes gives him the signal to stop. It is the trainer who does the counting.

One of the first symbols for a fraction came from the country where this man lived long ago.

WHY DO WE COUNT EGGS BY THE DOZEN?

You can divide a dozen eggs in several ways without cracking any shells. Twelve is divisible by two, three, four and six. Ten, however, is divisible only by two and five. That makes it easier to do business with numbers in groups of twelve than in groups of ten. People have known this for a long time.

About a thousand years ago in Europe King Charlemagne decided that the money in his country had to be divided into twelve units. In those days there were no quick and easy ways to do calculations with fractions. So it was important to have answers to money problems that would come out in whole numbers.

In old English money, twelve pence equalled one shilling. In the United States people still measure with a foot that is twelve inches long. And on clocks all over the world, twelve plus twelve hours equal one day.

WHO INVENTED FRACTIONS?

People used fractions long before they had any name for quantities such as one-half or one-third. Long ago, when a woman broke a piece of bread into two or three parts and shared it, she gave each member of her family a fraction of it. The English word *fraction* was actually made from an ancient word that meant *broken*.

Detective work has turned up a few facts about the invention of written fractions. Scientists know that one inventor lived in Babylonia in western Asia nearly five thousand years ago. They have found the ruins of an ancient city where he worked. They think he was a man because in ancient times women were seldom taught to read and write.

Nobody speaks the inventor's language now. But experts who study ancient writing have followed many clues and found out how to read the words he used.

The experts discovered that he was a bookkeeper. He wrote down records showing how much grain and other food people gave to the king. The records have lasted so long because they were written on soft clay tablets, and then the tablets were baked hard.

There is another clue in the shape of a number the bookkeeper made. It shows that someone brought the king a container only half full of grain. The fraction sign for ½ is a picture of a container cut in two like this:

Bookkeepers in neighboring countries borrowed the idea of fractions, but made up their own signs. Then, in far-distant places the idea was invented all over again. In ancient Greece a strange thing happened. Mathematicians refused to use fractions because they thought that whole numbers were too wonderful and important to be broken up. But Greek slaves learned to calculate with fractions. Slaves were the ones who had to do practical arithmetic, and fractions are very practical numbers.

Later, when traders began to do business all over the world, they had to know arithmetic. At last almost everyone who went to school learned arithmetic, and that meant learning fractions.

WHAT ARE DECIMALS?

It's easy to tell how much three magazines will cost if each one costs half a dollar. You can figure this out in your head. The answer is one dollar and a half. You write this $1.50. The dot is a decimal point and .50 is a decimal fraction meaning one-half.

Decimal fractions, or decimals for short, are a very useful invention. Scientists use them. Bookkeepers use them. So do waitresses and garage mechanics. You use them, too, to keep track of your dollars and cents.

WHO INVENTED DECIMALS?

In 1492, the same year that Columbus reached America, a man named Francesco Pellos used the decimal point to indicate decimal fractions. So far as we know, he was the first to do so. Instead of writing 2½, he wrote 2.5. Although he could write decimal fractions, Pellos did not understand how to calculate with them. The first person to do that was a German, Christoff Rudolff. In 1530 he used decimal fractions in a book about money. Rudolff put a bar instead of a point between the figures in his decimals. For ten and one-quarter, he wrote 10|25.

A few years later, a Frenchman named François Viète also began using decimals. Sometimes he wrote them with a bar and sometimes with a comma, like this: 10,25. He tried to persuade other people to use decimal fractions, but he didn't succeed, even though he was considered a very clever man. While Viète was alive, he was better known for his ability to figure out secret codes than for his work with decimals. He was so good at decoding the messages of foreign enemies that some people accused him of getting help from the devil.

The first person to write a book about decimal fractions was Simon Stevin, who lived in the Netherlands. In 1585 Stevin explained the rules for using decimals. Although his explanations were good, his way of writing decimals was hard to read. What we write as 189.375, Stevin wrote as 189, 1 3, 2 7, 3 5.

Others began to take up the idea of decimals. In 1617 John Napier wrote a book in which he used the decimal point the way we do today. But it was more than 150 years before most people began to use decimal fractions.

WHAT IS A ROUND NUMBER?

Suppose someone asked you, "What is the distance around the earth?" You might answer, "At the equator it is about twenty-five thousand miles." Your answer is a round number. The circumference of the earth at the equator is really 24,901½ miles, but twenty-five thousand miles is much easier to remember. When we express a number in hundreds or thousands, we say it is "rounded off."

Another number that is often rounded off is the speed of light. Light travels 299,792½ kilometers a second or 186,282 miles a second. These numbers are hard to remember. So people say the speed of light is about 300,000 kilometers per second or 186,000 miles per second.

If you need to know the exact speed of light or the exact circumference of the earth, you can always look it up in an encyclopedia.

WHAT IS A SQUARE NUMBER?

A square number is formed by multiplying a number by itself. Multiply two by two, for example, and you get four, which is a square number.

This is easy to understand if you use dots to represent a square number. First make a row of dots. Under each dot, make a column of dots. In each column, put the same number of dots as in the starting row. When you finish, the number of dots in all the rows will equal the square of the number of dots in one row. The smallest square number is easy to show. It is one, for $1 \times 1 = 1$. The next square number is four, shown by two dots in each row and two in each column. After that comes nine, then sixteen.

You can tell what number each diagram stands for by counting all the dots in the square. But a quicker way is to count the number of dots on a side and multiply that number by itself.

It is also possible to draw a rectangular number that is not a square. In such a number, the rows and the columns are different in length. Here is one of them:

You can tell the number it represents by counting all the dots. But the easier way is to multiply the number of dots in a column by the number of dots in a row.

ARE THERE ANY TRIANGULAR NUMBERS?

Yes, there are triangular numbers. To make a triangular number, start with one dot. Under it put a row of two dots. Then add a row of three dots. Continue this so that each row has one more dot than the row before it. When you finish, add up all the dots and you have the triangular number shown in this four-row diagram:

Count the dots in it and you will find that there are ten. One, plus two, plus three, plus four adds up to ten. And perhaps you think there's nothing special about that. The Greeks of ancient times thought differently.

A group of Greeks called the Pythagoreans thought triangular numbers were sacred. To the Pythagoreans one, two, three, and four represented fire, water, air, and earth. They thought these four things were the elements. Everything in the world was made up of them. And that included cats, houses, garbage — even people. The Pythagoreans used the four-row triangular number ten as a symbol of their beliefs. They called it the "holy fourfoldedness" and prayed to it, since it was the source of everything.

The Pythagoreans knew that square numbers and triangular numbers are related. If you add any triangular number to the next higher triangular number, you will have a square number. The diagrams below show this.

$1 + 3 \quad 3 + 6 \quad 6 + 10$

In modern mathematics, we often use square numbers. But triangular numbers are curiosities. Though they were so important to the Greeks, triangular numbers are of little use today.

DOES 1 + 1 ALWAYS EQUAL 2?

That depends on what number system you are using. The one that you use every day — zero through nine — is called the decimal system. When you use it, 1 + 1 = 2. But a system called the binary system is different. In this system, one plus one still equals two, but you write it 1 + 1 = 10. The binary numeral 10 has the same value as the decimal numeral 2.

The binary system uses only two symbols, zero and one. Here are the binary and decimal numbers from zero to twenty:

All the calculations that can be done with decimal numbers can be done with binary numbers. But look at twenty in the binary system. It is five digits long! Large binary numbers are very long, and that's one of the disadvantages of this system. When people do calculations with pencil and paper, they prefer the decimal system with its shorter numerals.

The binary system was invented several centuries ago. For hundreds of years it was just a curiosity. That changed when the electronic computer was invented. The binary system is convenient for computers because zero and one can be represented by the *on* and *off* positions of an electric switch. Computers work so fast that the length of large binary numbers is no problem for them. In just a few minutes they do calculations that would take people years and years to complete.

Binary:	0	1	10	11	100	101	110	111	1000	1001	1010
Decimal:	0	1	2	3	4	5	6	7	8	9	10
Binary:		1011	1100	1101	1110	1111	10000	10001	10010	10011	10100
Decimal:		11	12	13	14	15	16	17	18	19	20

DO ALL PEOPLE COUNT THE SAME WAY?

We use a counting system based on the number ten. We count from zero to nine with the symbols 0, 1, 2, 3, 4, 5, 6, 7, 8 and 9. Numbers after nine are written with combinations of these ten symbols. To go above ten we write 11, which means 10 plus 1.

Nowadays the base ten number system is used all over the world. But this was not always so. The people of Queensland, Australia, used to say "one, two, two-plus-one, two twos, much." "Much" was used for all numbers greater than four. There was little need for big numbers, for the Queensland people had very few possessions. They did not keep cattle or sheep that had to be counted.

In Central America, the Maya Indians led more complicated lives. Maya priests were star-watchers who kept accurate track of time. They invented a system of counting by twenties. The Maya wrote their numerals using dots, bars and a symbol for zero. Their numerals from one through twenty were:

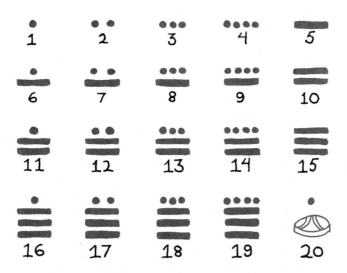

The largest number ever used as a base for a number system is sixty. The Babylonians invented that system thousands of years ago. It has one big advantage, especially for people who have trouble with fractions. Sixty can be divided evenly by two, three, four, five, six, ten, twelve, fifteen, and thirty. If you work with a base of sixty, you can often avoid fractions.

Traces of the Babylonian number system still appear on our clocks. One hour equals sixty minutes, and one minute equals sixty seconds.

How many numerals can you find on this picture drawn by Maya Indians? Hint: The dots and bars can go up and down as well as crosswise.

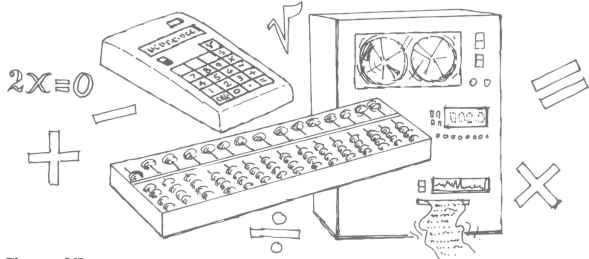

Chapter VI

Figuring it Out

1. How can you add without writing down numbers?
2. Which is faster: an abacus or an adding machine?
3. Where did the signs for "plus" and "minus" come from?
4. Where did the sign for "multiply" come from?
5. Where did the sign for "divide" come from?
6. Where did the sign for "equal" come from?
7. Can you multiply just by adding?
8. How can 10 + 5 = 3?
9. What is the difference between arithmetic and algebra?
10. What does x mean?
11. Who invented the computer?
12. What is a computer program?
13. What is a bug in a computer program?
14. Can a computer remember all the numbers in a phone book?
15. How does a computer's memory work?
16. Can a machine play chess?
17. Can a computer make you richer?
18. How does a pocket calculator work?
19. Is there any difference between a calculator and a computer?
20. Can you make a computer cheat?
21. Can computers think?
22. What has digits but no hands?
23. What was the first digital computer?
24. Why do the numbers on a pocket calculator look so strange?

57

HOW CAN YOU ADD WITHOUT WRITING DOWN NUMBERS?

If the numbers are small, it is easy to add without writing anything down. Just do the problem on your fingers. But when the numbers get big, you will need a better way to keep track of them. One way is with an abacus.

When you use an abacus, you count beads instead of fingers. The beads on an abacus slide up and down on thin rods. A center bar separates the abacus into an upper part and a lower part.

The beads on the rod farthest to the right are used to make the numbers from 0 to 9. To form 1, slide one bead from the lower group up against the center bar. To form 2, slide 2 lower beads up. To form 3, slide 3 up, and for 4, slide all 4 of the lower beads up.

To make 5, leave all the lower beads at the bottom, and slide the upper bead down against the bar. The pictures below show how to make the numbers from 5 to 9.

The other rods are used to make numbers larger than 9. The pictures show how to make 10, 11, 13, and 313.

How can you add with an abacus? To add 313 and 11, first make the pattern for 313. Then add 11 to 313 by sliding up one bead on the rod on the right and one bead on the second rod. The new bead arrangement is the pattern for 324, which is 313 + 11.

The abacus was invented long before Arabic numerals. It has disappeared in many places, but it still is used in China, the Soviet Union and Japan. Many people in those countries prefer it to pencil and paper. They find it is easier and faster to do arithmetic with an abacus.

WHICH IS FASTER: AN ABACUS OR AN ADDING MACHINE?

On November 11, 1946, a trained Japanese abacus operator and an American soldier using an adding machine held a contest in Tokyo, Japan. Which one could do arithmetic problems faster? Many of the three thousand people who watched the contest were Americans. When they saw how quickly the abacus operator could push the beads, they thought of a perfect nickname for him: "The Hands." Moving the beads at incredible speeds, he found the answers to most of the contest problems long before the soldier finished his calculations on the adding machine. And "The Hands" made fewer mistakes.

How would an abacus do in a contest with a pocket calculator? Addition and subtraction problems can probably be done faster on an abacus. An experienced abacus operator can push the beads much faster than most people can push calculator buttons. For multiplication and division problems a pocket calculator is probably faster. Multiplication and division are complicated to do on an abacus, but on a calculator, the answer appears after only one push of a button.

WHERE DID THE SIGNS FOR "PLUS" AND "MINUS" COME FROM?

Plus and minus signs are short cuts. People invented them long ago to avoid writing out "add the second number to the first number." Or "subtract the second number from the first number."

In ancient Egypt, bookkeepers used two funny little short-cut signs. A pair of legs walking to the right like this ∧ meant "add." Legs going to the left like this ∧ meant "subtract."

Later, people who wrote in the Latin language invented a short cut when they got tired of spelling out their word for "and." They made the sign + instead. Nobody knows why they used it, except that it is easy to make. Bookkeepers and merchants in Europe then adopted the sign. We still say "two and two make four," but we write it with the short-cut sign +.

The sign for minus probably began about 500 years ago as a short cut for the word *minus*. Bookkeepers first wrote it m^s. Then $\overline{m}$. Finally they left out the m and used only the straight line. So 8 m^s 4 became 8 − 4.

WHERE DID THE SIGN FOR "MULTIPLY" COME FROM?

For hundreds of years a cross like this X had special meaning for people in the Christian part of the world. They believed that St. Andrew, one of Jesus' disciples, was tied to that kind of cross and left to die.

More than 300 years ago an Englishman named William Oughtred wrote out multiplication problems using a St. Andrew's cross between the numbers to be multiplied: 4 × 2. Soon most writers of arithmetic books in Europe were using Oughtred's sign.

But then mathematicians began to use the letter x in problems. The x stood for an unknown number. They complained that × could be confused with x. Many mathematicians decided to use a dot as a sign for multiplication.

Today you can find the dot in some math books and the cross in others. But watch out for the position of the dot. In England it is placed low between numbers, like a period, so that 2 . 2 = 4. In the United States the dot is raised a little, so that 2·2 = 4.

About the same time that William Oughtred used × for multiplication, another writer had a different idea. He made his sign like this: *. The star-shaped sign could not be confused with x, but it didn't catch on for 300 years. Today people who work with computers use a six-pointed star to indicate "multiply."

WHERE DID THE SIGN FOR "DIVIDE" COME FROM?

The sign for "divide" has been a troublemaker for hundreds of years. It began peacefully when an Arab writer 800 years ago explained how to show one thing divided into two parts. He said, "Write it like this: ½."

Later there were people who decided to put dots above and below the line to show division, like this: 1 ÷ 2. Others turned the sign so that the bar was vertical ·|· and still others used two dots : that look like a modern punctuation mark. One writer even suggested 2)4(2. This meant 4 divided by 2 equals 2.

For some reason mathematicians got very stubborn about the whole business of division signs. They still don't agree. So in England and in the United States and other English-speaking countries ÷ means "divided by." On the continent of Europe and in Central and South America the two dots : are the sign for division.

WHERE DID THE SIGN FOR "EQUAL" COME FROM?

Try to think of the most nearly equal things in the world. Peas in a pod? Twins? How about twin lines? That is, lines that run parallel. Parallel lines are always an equal distance apart. Probably it was this idea that Robert Recorde had in mind when he invented the sign for "equal" more than 400 years ago in England.

Many mathematicians soon adopted Recorde's = sign. Others preferred inventions of their own:

$$\not\equiv \quad \sqcup \quad \sim$$

But in the end Recorde's twin lines won out. They are among the few signs that almost all European and English-speaking mathematicians agree on.

Adam Riese, a famous German writer of arithmetic textbooks. This portrait is from one of his books printed in 1550. Notice the multiplication sign.

CAN YOU MULTIPLY JUST BY ADDING?

In the year 1614 John Napier, a Scotsman, said "Yes! You can multiply by adding." In those days there were no pocket calculators and no computers. Long, tedious arithmetic problems had to be done with pencil and paper. Napier wanted to make these calculations simpler, and so he invented a system of numbers called logarithms.

Napier had discovered that every number has a special companion-number — a logarithm. And he found that you can multiply two numbers just by adding their logarithms together.

Some numbers have logarithms which are extremely easy to remember. For example, 1,000,000. One million is a 1 followed by 6 zeroes. Its logarithm is 6. One hundred is a 1 followed by 2 zeroes, so its logarithm is 2. The logarithm of one billion (1,000,000,000) is 9.

No one who uses logarithms likes to write out the whole word. It's just too long. So instead of writing

logarithm of 1,000 = 3

most people write

log 1,000 = 3

Now, suppose you want to multiply 100 times 1,000 using logarithms. Here's what you do: First, find the logarithms of 100 and 1000. If you don't remember them, you can look them up in a list that Napier made.

log 100 = 2
log 1,000 = 3

Now *add* the logarithms together.

2 + 3 = 5.

The number 5 is a logarithm, too. When you look it up in Napier's list, you find that 5 is the logarithm of 100,000. And 100,000 is exactly 100 times 1,000! In other words, if you want to multiply two numbers, just add their logarithms.

Napier didn't stop with the easy logarithms such as log 10, log 100, and log 1,000. He figured out how to get the log for any number at all. This was a painstaking job. But after he had arranged the logs in his long list, which is called a log table, other people could use the list to do short-cut multiplication.

Today logarithms are less important than they were in Napier's time. Using a calculator, you can do the most difficult multiplication problems just by pushing a few buttons. But logarithms are still necessary. Scientists and mathematicians use them in solving certain special problems.

HOW CAN 10 + 5 = 3?

The train ride from Boston to New York takes 5 hours. If you leave Boston at 10 A.M., you will arrive in New York at 3 in the afternoon, so 10 + 5 = 3. This special kind of arithmetic is called clock arithmetic.

How much is 8 + 12? Suppose you take a 12-hour train ride that starts at 8 A.M. The train arrives at 8 P.M., so 8 + 12 = 8. Of course, 9 + 12 = 9 and 4 + 12 = 4. For clocks, adding 12 is just like adding 0.

Does clock arithmetic work for bigger numbers? How much is 17 + 30? The number 17 equals one complete trip around the clock face, plus 5 hours. The number 30 equals two complete trips around the clock face, plus 6 hours. So 17 + 30 equals three complete trips plus 11 hours, and we write
$$17 + 30 = 11.$$
What about 17 + 31? This equals four complete trips around the clock, plus 0 hours, so
$$17 + 31 = 0!$$

WHAT IS THE DIFFERENCE BETWEEN ARITHMETIC AND ALGEBRA?

Suppose you are in a bakery, and you are hungry. You want to buy two rolls and a doughnut. The rolls are 10¢ each and the doughnut is 18¢. How much money will you need? You can answer this question with arithmetic. Arithmetic is the system for adding, subtracting, multiplying, and dividing numbers. Questions in arithmetic are called problems. Your math problem at the bakery is to add 10 and 10 and 18. Do you have 38 cents? Your problem is solved.

An arithmetic problem can be made of many numbers. But you must know all the numbers to get an answer. You might want to buy a ten-cent roll and an eighteen-cent doughnut and a cupcake. If you did not know the price of the cupcake, you would not have an arithmetic problem.

When you don't know all the numbers in a problem, algebra is the system for solving the problem. You use letters to stand for the unknown numbers. The rules of arithmetic are also true in algebra. For example, 4×6 is the same number as $6 + 6 + 6 + 6$. Both equal 24. In algebra, $4 \times a$ equals $a + a + a + a$. The letter a can stand for any number.

WHAT DOES X MEAN?

Here's a riddle:
 Rachel is twice as old as Josh.
 Rachel is 14.
 How old is Josh?

Riddles ask you to solve a mystery. In this one you have to find a mystery amount: Josh's age. Algebra problems are just a kind of riddle, and when people are solving riddles they give the mystery amount a special name: x.

Using x makes writing algebra easy. If you call Josh's age "x," then twice Josh's age is just twice x, or $2x$. The riddle says that twice Josh's age equals Rachel's age, and Rachel is 14. And so we can write down the whole riddle in one line of algebra:

$$2x = 14$$

The rules of algebra say that we can divide both sides of the equal sign by the same number. To solve this riddle we divide both sides by the number 2. When we do, we get:

$$x = 7$$

and this answers the riddle. Josh is 7 years old.

Why is the mystery amount called x? Four hundred years ago people in Europe were writing a great many algebra books. But one author didn't always know what another was doing. The mail was slow and unreliable, and of course there were no telephones. Each author had his own special way of writing algebra problems. When authors in Italy talked about the mystery amount in a problem, they sometimes called it *cosa,* the Italian word for *thing.* In Germany, they often used the German word for number, *zahl.* Some people used *res,* the Latin word for *thing:* 2 *res* = 14.

Then, in 1637, a very important book was published in France. In it the author, René Descartes, used the letters x, y, and z to stand for the unknown amounts in his math problems. But toward the end of the book, the letters y and z appeared less and less often. Why was this?

In those days, type for printing was expensive and hard to get. So printers had many more pieces of type for common letters than for uncommon letters. Most books in France were written in French or Latin, and words in these languages have many more x's in them than y's or z's. The French printers who set the type for Descartes' book used up all their y's and z's in the early chapters. They probably told Descartes to finish his book using x as often as possible.

Descartes' book was very popular, and using x to stand for an unknown amount caught on. So you won't find *cosa* or *zahl* or *res* in math books today.

A modern computer

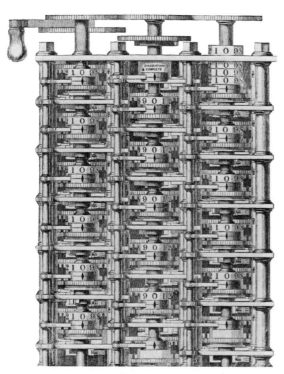

Charles Babbage's Difference Engine

WHO INVENTED THE COMPUTER?

June 14, 1822, was an important day for Charles Babbage. On that day the 30-year-old Englishman showed a group of scientists his plans for a computer. Babbage had already built a small working model which did simple addition problems and a few other easy computations. The new machine, which he called a "difference engine," would be much larger. No one had ever heard of anything like it. It would be able to do arithmetic, calculate complicated mathematical tables, and automatically set the type which would print the tables on paper.

Babbage's plan amazed the scientists who came to hear his talk. The British government was impressed, too. It agreed to pay part of the cost of building his machine.

At first, he thought the project would take only three years. But he found that the parts for his machine were very hard to make. After ten years of work, Babbage gave up. He had a new plan. He would build an even more amazing device which he called an "analytical engine."

This new engine was remarkably similar to a modern computer. It was to have nearly as much calculating power as today's machines. It would even have a computer memory. Of course, his machine would be much slower than modern computers, which work by electricity. The analytical engine was to be completely mechanical, with steam power driving all its gears, shafts, and wheels.

Like the difference engine, the analytical engine was never built. It was just too complicated. People weren't skillful enough to make all the delicate parts it needed. And Babbage ran out of money. At one point, he designed a machine to play tic-tac-toe. He thought he could help pay for the analytical engine by giving performances with his tic-tac-toe machine. But he gave up on this idea, too.

Babbage died a disappointed man. His dreams of a calculating engine never came true in his lifetime. It was almost a century before anyone built a machine that could equal the ones he planned but could not finish.

WHAT IS A COMPUTER PROGRAM?

Suppose your mother gives you the following list of errands to do:

1. Get the $20 bill on the table.
2. Go to the post office and buy some stamps.
3. If the fish store is open, pick up the fish.
4. If it's closed, come right home.
5. If you got the fish, stop at the grocery store and get two lemons to make a sauce for the fish.
6. Come home.

This list of instructions is a lot like another list of instructions — a computer program. In fact, that's just what a computer program is — a list of "errands" you want a computer to do.

Of course, computers can't buy stamps or fish. Computer instructions tell computers to do such things as add and subtract numbers. An instruction like this:

ADD '15',X

tells the computer to add 15 to a number which is called X.

Still, your mother's list and a computer program have many things in common. Your mother tells you how to start and finish your errands. You start by getting the money off the table, and you finish by coming home. If you write a computer program, you make a list of instructions on a special typewriter. To tell the computer when to start, you type out "START." To finish, you write "END."

Your mother has numbered her list because she wants you to do the errands in order. If you did them out of order, you might go to the post office without picking up the money to pay for the stamps.

Computer instructions in a program are numbered, too. And the computer usually follows the list of instructions in order. Sometimes, though, the computer comes to a special instruction which may allow it to skip a few instructions. But this also happens with your errand list. If the fish store is closed, you are supposed to come right home and not bother with those lemons.

WHAT IS A BUG IN A COMPUTER PROGRAM?

"ERROR IN LINE 40"

When you see this message on a computer television screen, you know exactly where to look for the "bug" in your program. Bugs, of course, are mistakes. You must *debug* your program — correct your mistakes — before the computer can give you the answer to a problem.

When the computer finds a bug in your program, the mistake is usually easy to correct. "ERROR IN LINE 40" often means that a word in that line has been misspelled. For example, line 40 of your program might be

ADE '3',X

instead of

ADD '3',X.

To correct the mistake, you type:

ERASE LINE 40.

Line 40 will disappear.
Then you type

ADD '3',X.

This puts the correct instruction into line 40 of your program.

If the misspelling in line 40 was your only mistake, then your program is ready to go. Just type

START

and in an instant the computer will follow all the instructions in your program and write the answer on the computer's television screen.

65

CAN A COMPUTER REMEMBER ALL THE NUMBERS IN A PHONE BOOK?

Computers are very good at remembering all sorts of things. In fact, an important part of a computer is called *memory*. It's not the same as your memory, of course, since a computer is a machine.

Computers can remember telephone books without any trouble at all. They can also remember lots of train schedules, photographs, and music.

Suppose you have your own computer. You can make it memorize the phone numbers of all your friends. First, you will have to write a phone-book program, a list of instructions for the computer that tells how you want the phone book to work. Your program will explain to the computer how to look for each friend's name and number in its memory. And the program will tell the computer how to write the correct phone number on the computer's television screen.

After your phone-book program is written, how will it work? First, you sit at the computer terminal and type your friend's name, like this:

ERICA LOGAN.

Her name will instantly appear on the computer's television screen. A split second later her phone number will appear on the screen next to her name, like this:

ERICA LOGAN . . . 441-8868

What happens if you go to the terminal and type:

ABRAHAM LINCOLN

If you don't have a friend named Abraham Lincoln, the computer won't have this name in its memory. When the computer reads a name it has never seen before, it follows special instructions which you have included in the phone-book program. Your instructions to the computer might be to write on the screen:

NO LISTING

or:

TRY AGAIN

You might even direct the computer to write:

WHO IS THIS GUY?
I NEVER HEARD OF HIM!

In fact, you can program the computer to say just about anything. After all, it's your phone book!

HOW DOES A COMPUTER'S MEMORY WORK?

Suppose you are solving a problem on a computer, and you want to keep track of a long list of numbers or words. You can store the list in the computer's memory. Later, when you need the list, you can have the computer recall it.

A computer's memory is made up of rows and rows of tiny electronic switches. These switches are a lot like ordinary light switches except that they work much faster. They will go on and off more than a half million times in a single second. A big computer may have as many as 64 switches in a row, and perhaps hundreds of thousands of separate rows.

A computer stores a word in its memory by forming special patterns of "on" switches and "off" switches. Each on-off pattern in a row has a meaning — often a letter or a number. The pattern for the letter A is:

off-on-off-off-off-on

And here is the pattern for the number 1:

off-off-off-off-off-on

If you store information in a computer's memory, how will you find it when you need it? Each row that can hold a pattern in storage is numbered, just in the way houses are numbered on a city street. In fact, a row's number is called its *address*.

Suppose the on-off pattern for A is stored at address 1000, and the pattern for 1 is stored at address 1001. Now suppose you want to recall A 1. You go to the computer terminal keyboard and type the instruction:

PRINT MESSAGE AT ADDRESSES 1000 AND 1001

Instantly the message "A 1" will flash on the terminal television screen.

CAN A MACHINE PLAY CHESS?

In 1770, a strange machine was put on display in the city of Vienna, Austria. The machine was a robot — a mechanical man — dressed as a Turkish sultan. The sultan sat behind a large box with a chessboard on top. Baron Wolfgang von Kempelen, who designed it, called it "The Turk." He said that The Turk could beat anyone at chess.

The Baron amazed audiences with exhibitions of his chess-playing machine. People were very suspicious, of course. And so, before every performance, the Baron would open the box and show the complicated machinery inside. Then the game would begin. And the Turk almost always won.

The Turk was a terrific chess player. But it wasn't a machine at all. It was a fraud. The real chess player was a midget hidden in the box. He was so short that he could keep out of sight even when the Baron opened the box before performances. During exhibitions the midget moved The Turk's chessmen, although nobody knows just how. Most people think the midget worked the Turk's hands and arms mechanically from inside the box. The Baron and later owners of The Turk always hid master chess players in the box, and so The Turk was almost unbeatable.

The only machines that can really play chess are electronic computers. A complicated computer program directs the computer as it plays. Computers are improving at chess. They do well against average players, but good players have little trouble beating them — at least, so far. Some computer scientists have made a bet that in a few years there will be a computer program that can play chess as well as almost anyone in the world.

"The Turk" — a "chess-playing" machine.

cause three months equals one-quarter of a year. At the end of the year you will have $105.09. That is a little more than you would get if the bank had paid simple interest of 5%.

It isn't hard for someone in the bank to figure out interest compounded quarterly, and it only has to be done four times a year. But suppose the bank wanted to pay interest compounded every day. Somebody in the bank would have to do long, tedious computations, then check the amounts that would have to be paid to everyone who has money in the bank. That would have been much too hard a job in the days before computers were invented. But a computer can do arithmetic at lightning speed. If you put a lot of money in a bank, and if the bank pays you interest compounded daily, the computer really can make you richer. A computer works so fast that it can calculate interest compounded every hour, every minute — even every second.

CAN A COMPUTER MAKE YOU RICHER?

Suppose you have $100 and you put it in a certain savings bank. At the end of a year that bank will tell you you have $105 in your account. The extra $5 is interest. It is called simple interest.

There is also another kind of interest called compound interest which some banks pay. This is how one kind of compound interest works:

At the end of three months the bank will add one-fourth of the year's interest to your $100. Now you will have $101.25. Three months later the bank will pay you one-fourth of a year's interest on $101.25. Now you will have $102.51½. Three months later the bank will repeat this same process, and again every three months afterward.

This kind of interest, which is paid every three months, is said to be compounded quarterly, be-

Starting with $100

interest the first day is
.05 ÷ 365 × 100

interest the second day is
.05 ÷ 365 × (.05 ÷ 365 × 100 + 100)

and

interest the third day is
.05 ÷ 365 × [.05 ÷ 365 × (.05 ÷ 365 × 100 + 100) + .05 ÷ 365 × 100 + 100]

The tedious, time-consuming way of calculating interest.

HOW DOES A POCKET CALCULATOR WORK?

Suppose you want to solve the problem 3 + 4 on your calculator. You push the "3" button, then the "+" button, then the "4" button, then the "=" button. And presto! there in the window is the answer — "7." How does it get there so fast?

Inside the calculator are many rows of tiny electronic switches. As you push the "3" button, a signal is sent to one of the rows. The signal turns on some of the switches in that row. The on-off arrangement of the switches makes a special pattern:

off-off-off-on-on

This is the pattern for the number 3.

The "+" button turns on a special switch. When this switch is on, all the signals that reach it are sent to a part called the adder.

Pushing the "4" button sends a signal to a new row and turns on the switches that make the pattern for the number 4. This is:

off-off-on-off-off

The "=" button sends out the most complicated signal of all. The signal goes to the adder switch and then to the adder itself. The adder sends out a new signal to the rows holding the "3" and "4" patterns, and the signal changes when it reaches them. It becomes a new signal, turning on a new pattern of switches in a new row, the answer row. From there, signals go to tiny light bulbs in the window. A 7 lights up about 1/20 of a second after you push the "=" button. And that's much less time than it takes you to say 3 + 4 = 7.

IS THERE ANY DIFFERENCE BETWEEN A CALCULATOR AND A COMPUTER?

A calculator and a computer are very much alike inside. Both have memories and special parts called adders. Although computers are much bigger and faster than calculators, there is little difference between them electronically.

Of course, calculators and computers are machines. They can do nothing at all until a person tells them what to do.

When you give instructions to a calculator, you push buttons marked +, −, ×, ÷. If these are the only instruction buttons, then you can only add, subtract, multiply, and divide.

Computers are different. If you use a special typewriter, you can give a computer a long list of instructions called a program. The computer can store this list for you in its memory. It can carry out the list of instructions any time you want, as often as you want. The instructions in the list might be + or ×. There might be others, too. You can tell the computer to back up five steps on the list and start again, or skip to some new instruction far down the list.

Because a computer can go back and forth on the instruction list, you can make it do far more difficult things than solving arithmetic problems. You can program a computer to land a spaceship on Mars, or play a game of checkers. You can even get a computer to read a book out loud!

CAN YOU MAKE A COMPUTER CHEAT?

A dishonest computer programmer once got rich quick using a very simple plan. The programmer worked at a bank, where he was in charge of the computer programs that kept track of the money people had in savings accounts. One day he made some changes in those programs: he added a few extra instructions for the computer.

One list of new instructions told the computer to subtract one cent every month from every savings account at the bank. Another new list of instructions told the computer to put all those cents into a special account the programmer had set up for himself.

If a customer had $100 in his account, the bank should have given him an interest payment of 46¢ at the end of the month. But now he only got 45¢, because the programmer's extra instructions subtracted one cent from the account.

None of the bank's customers noticed. No one missed the cent that disappeared from each monthly interest payment. After all, people thought, a bank can never be wrong.

The programmer got richer and richer. And it was a long time before he was caught.

The "penny pincher" in the story is not the only computer thief in the world. Altogether, computer bandits stole about $300,000,000 in the United States in 1975!

CAN COMPUTERS THINK?

Computers are actually very dumb. Without a special list of instructions, called a program, they can't do the simplest problem. Somebody must type out the list of instructions on a special typewriter which puts the program in the computer's memory. Only then can a computer begin to work.

After computers have been told what to do, they can sometimes do things that seem like thinking. For example, there is a program that tells a computer how to play checkers. And the computer almost never loses. In this checkers program, a human player writes out a list of rules that the computer uses when it plays. One rule is: Don't move checkers out of your back row until it's absolutely necessary. All these rules put together are called a strategy. Some strategies are better than others. With a good one, a computer can plan clever moves very far ahead. Most people can't do this, and so the computer almost always wins.

Is this thinking? Some scientists say that it is. But most people believe that thinking is something that only human beings can do. Machines can quickly solve problems it would take a person years to work out by hand. But no computer can use human language to carry on a real conversation. A computer can't know when a story is sad — and it could never understand a joke.

| 60 | 800 | 30,000 | 50,000 | 10 |

WHAT HAS DIGITS BUT NO HANDS?

Everybody has digits — "digit" is another word for finger or toe. Everybody has hands, too. Well, who — or what — has digits but no hands? This question has two answers, one ancient and one modern.

When the ancient Egyptians told time with a sundial, they were using a clock that had digits but no hands. A sundial is a flat disc with a post in the middle. Digits — that is, numerals — standing for the hours of the day are written around the border of the disc. They look like the numerals on the face of a clock. When a sundial is in the sun, the post casts a shadow that points to a numeral on the face. As the sun moves across the sky, the shadow moves from numeral to numeral around the face. So a sundial is a clock without hands. Instead of hands, it uses the shadow of the sun to show the time.

There is another kind of clock with digits but no hands. This kind is very new. It is called a digital clock. It has a plain dark face, and the time of day appears in bright, glowing numerals that look like this:

8:15:31

This means that the time is 15 minutes and 31 seconds after 8 o'clock. The numerals change every second. On some clocks they may change every minute.

Inside a digital clock there is a tiny lump of a mineral called quartz. When the clock is running, an electric current makes the lump vibrate. The lump vibrates 32,768 times every second. A tiny computer inside the clock keeps track of the vibrations. When the computer has counted 32,768 vibrations, it adds one second to the time shown on the face of the clock.

WHAT WAS THE FIRST DIGITAL COMPUTER?

The number 6,781,540 has 7 digits. A digit is any one of the numerals from 0 through 9.

The word digit also has another meaning. Your fingers and toes are digits. Ever since the day when some person first wanted to check the number of fish that had been caught or nuts that had been gathered, people have been counting on their fingers and toes.

Finger-counting did not stop when written numbers were invented. Through the ages people worked out different ways of showing numbers by the positions of their fingers. With fingers alone they could show numbers up to 9,000. Then, by adding gestures — hand on head, hand on heart or hip or stomach — they could go on up to a million. They could do even more. A skillful person could learn to multiply and divide big numbers using only his fingers.

To this day some people in the world do not know how to write numbers or do written arithmetic. But their finger-computers still work.

Finger-computers really aren't so different from modern digital computers. In a modern computer the "fingers" are tiny electric switches that open and close. Opened and closed switches make patterns that stand for numbers. This is similar to the number pattern a person can make with straight and bent fingers.

WHY DO THE NUMBERS ON A POCKET CALCULATOR LOOK SO STRANGE?

When you push the "8" button on your calculator, the number 8 instantly appears in the space across the top. This space, which looks like a slim dark window, is called the calculator display. The 8 you see on the display is not like the rounded figure 8 you make when you write numbers with a pencil. It is a double-square 8, and it has this shape:

The other calculator numbers have square corners, too. They look like this:

There is a good reason why the numbers have these shapes. Each number can be formed by straight lines. And each of the straight lines can be made by turning on a tiny light bulb behind the calculator display window. With a total of only seven bulbs, any digit from 0 through 9 can be made to appear on the display.

When you push the "8" button, all seven bulbs light up. Push the "3" button, and only five of the bulbs light:

In a calculator display there are usually 8 or 9 places where a number can appear. And behind each of these places is a double square made up of seven bulbs.

Now try this on your calculator: push the buttons to make the number 5318804 on the display.

Next, turn the calculator upside-down. You will see the word "hobbies"! Some calculator numbers, when you see them upside-down, look like letters.

By turning your calculator right-side up, then upside-down, you can make up all sorts of riddles, like this one:

Question: What's the difference between happiness and sadness?

Answer: 49,373.

This is how you get the answer:

Bliss — another word for happiness — is 55,178 turned upside down.

Sobs — you make sobs when you are sad, and sobs is 5,805 turned upside-down.

Now, when you subtract 5,805 from 55,178, you get 49,373. So 49,373 is the difference between happiness and sadness.

Chapter VII

So Big, So Small

1. When was a foot not a foot?
2. How can the ocean's depth be measured?
3. How far is it to the bottom of the sea?
4. Can a fat man reduce by going to the top of a mountain?
5. Which weighs more: a pound of feathers or a pound of gold?
6. What did these measures measure?
7. And what about these?
8. What is a hairbreadth escape?
9. Why didn't the United States adopt the metric system when other countries did?
10. How do we know the length of a meter?
11. Why do thermometers have different scales?
12. How do we know how many calories are in a hamburger?

WHEN WAS A FOOT NOT A FOOT?

In ancient times people measured things by parts of the body. The distance from the elbow to the tip of the middle finger was a cubit. If you put one foot down on the ground, then put the other just in front of it, the two feet added up to about a cubit. A little more or a little less didn't matter very much. Measurements did not have to be very exact when you were making a hut of mud or animal skins.

But suppose that a king decided he wanted a large building made of brick or stone. Now the problem was different. Big buildings could be made solid only if the bricks or stones were precisely measured, so that they fitted together exactly. Whose foot was going to be the one the builders measured by? The king's, of course.

Of course, the king didn't step on every brick to make sure it was the right size. Instead, the builders cut a measuring stick just the length of his foot. That would be half a cubit. We know that was so in a place called Lagash, in Mesopotamia, about four thousand years ago. The monarch, whose name was Gudea, had artists carve statues that show him with a foot rule in his lap. The rule is about 10½ inches long by English and United States measurement.

Ancient Greek people may have had bigger feet than those of Gudea. Their foot measure was a little more than the present-day 12 inches. Other people used foot rules that were about 11 inches. In ancient Germany a foot was about 13 inches.

Every country, sometimes every town, in the old days made up its own length for a foot. This could be annoying. Suppose a trader came from a place where a foot measure was 11 inches long. Now suppose he brought some cloth to your town where everybody used a measure 12 inches long. If you needed to buy 4 feet of cloth, the trader would measure it by his foot rule. But since his rule was an inch shorter than the one that people in your town used, you would get 4 inches less cloth than you needed.

At last, about 700 years ago, King Edward I of England made a law that set an exact measurement. He had some iron rods cut, all the same length. Each was now the official three-foot rule, the first official yardstick. The king sent the rods to sheriffs all over England. It was their job to see that no merchant cheated customers by using a yardstick that was a little bit short.

HOW CAN THE OCEAN'S DEPTH BE MEASURED?

In the days before ships made long voyages on the deep sea, sailors constantly measured the water's depth as they sailed along near land. They wanted to avoid places where their ships might hit against rocks or get stuck because the water was too shallow.

To do the measuring, sailors used a rope fastened to a heavy weight that sank quickly to the sea bottom. All along the rope were markers made of leather or cloth. Even in the dark a sailor could tell by feeling how far up on the rope the water had reached. A marker made of two strips of leather meant that the weight had sunk two fathoms before it reached bottom. A fathom was an ancient measurement equal to the distance from the tips of a man's outstretched fingers, across his chest, to the opposite fingertips.

Viking sailors measured by fathoms. So did men who worked on boats on the Mississippi River a hundred years ago. If the water was two fathoms deep, a Mississippi boatman would sing out "By the mark twain." (Twain is an old word for two.) One riverboatman, named Samuel Clemens, later wrote *Tom Sawyer* and *Huckleberry Finn*. But instead of using his own name, he called himself Mark Twain.

Today a fathom is a more precise measurement — 6 feet or 1.829 meters. But as a rule, deep-water sailors do not need a marked rope for measuring. Instead, big ships carry an automatic device, sometimes called a Fathometer, which measures the distance to the sea bottom. Electric current in the Fathometer creates sound waves which travel downward through the water. In very cold sea water they go 4,742 feet per second. When they reach bottom, the sound waves bounce back — just the way sound waves in the air echo when you shout toward a cliff or brick building.

Like your eardrums, the Fathometer can pick up the echo. The Fathometer can also time the echo. Suppose it takes the sound waves two seconds to make the round trip to the bottom and back — one second down and one up. That means the ocean is 4,742 feet deep at that spot.

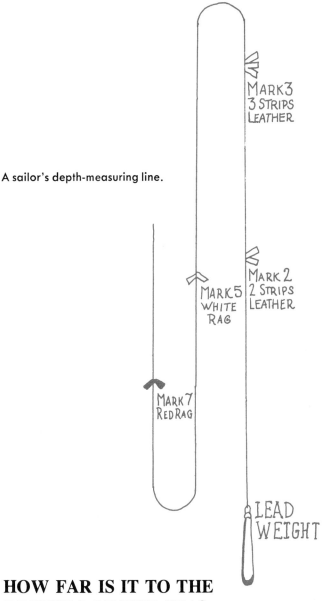

A sailor's depth-measuring line.

HOW FAR IS IT TO THE BOTTOM OF THE SEA?

The floor of the ocean is a good deal like the earth's dry land. In some places it is low and flat; in others there are high plains and tall mountains. Deep canyons called trenches cut through it here and there. Usually the water is rather shallow near continents or close to large islands. Away from the land it gets deeper. So far as anyone knows, the sea is deepest at a place called the Mindinao Deep near the Philippine Islands in the Pacific Ocean. There the sea bottom is 11,516 meters below the surface. That is more than 7 miles — 11½ kilometers — straight down!

Weight is different from mass. A man's weight is the pull that gravity exerts on his body. The pull of gravity is strongest at the center of the earth. It is less strong at sea level. It is still less on a mountaintop. And so the farther a man gets from the center of the earth the less he weighs, although his mass is the same.

Now that you know there is a difference between weight and mass, you can go right on saying weight when you really mean mass. Everybody else does in ordinary life.

WHICH WEIGHS MORE — A POUND OF FEATHERS OR A POUND OF GOLD?

After the year 1878 a pound of feathers weighed more than a pound of gold in the United States — but not in England! As usual, the story begins a long time ago. More than a thousand years ago the smallest unit for measuring weight in England was a grain of wheat or barley. Kings in those days used grains when they were weighing their treasure. Exactly 480 grains equaled an ounce of gold, and 12 ounces — that is, 5,760 grains — made a pound. For some reason, this was called a troy pound.

However, there was another kind of pound used for measuring ordinary things like feathers. It was called an avoirdupois pound. There were 437½ grains in an avoirdupois ounce, and 16 ounces — that is, 7,000 grains — made a pound. So a pound of feathers weighed 1,240 grains more than a pound of gold.

Until 1878 both kinds of pound were used in England. In that year the government discontinued the troy pound. The United States Government did not. So, in the United States, a pound of gold still weighed only 12 ounces and a pound of feathers weighed 16 ounces.

One more complication: England kept troy ounces as a measure even after the troy pound was abolished. So, in both England and the United States, an ounce of gold weighs more than an ounce of feathers.

CAN A FAT MAN REDUCE BY GOING TO THE TOP OF A MOUNTAIN?

The answer depends on how he gets there. If he climbs all the way up to 20,000 feet from sea level, he may lose a few pounds of fat, provided he doesn't eat much on the way.

But suppose a man who weighs 200 pounds at sea level takes a helicopter. And suppose he lands on a mountaintop 20,000 feet high. Now, even without any exercise, he weighs only 199 pounds when he gets on the scales. Did he really reduce? The answer is "No." The quantity of fat and flesh and bones — all the stuff in his body — did not change. Scientists say his *mass* did not change.

WHAT DID THESE MEASURES MEASURE?

At one time or another all the words in this list have been names for measures of various things that people bought and sold. Try guessing, and then check the answers.

anker	cab	ell	nail	shekel
bodge	catty	firkin	perch	shid
bow	clove	hide	rotl	to
butt	cran	log	seer	tum

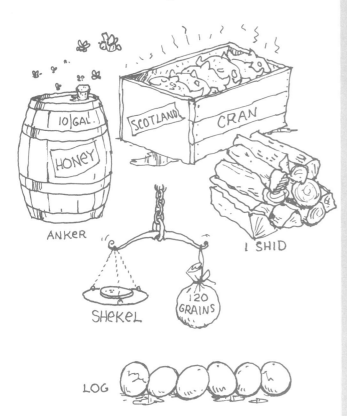

This is what they measured:

anker	10 gallons of honey (Holland and Germany)
bodge	½ gallon of liquid (England)
bow	about 8 feet of length (ancient India)
butt	a big barrel of wine (Greece)
cab	a measure of grain (Ancient Israel)
catty	a sack of rice (old China)
clove	8 pounds of wool or cheese (England)
cran	37½ gallons of fresh herring (Scotland)
ell	a measure of about 1 meter (Denmark and other European countries)
firkin	56 pounds of butter (England)
hide	100 acres of land (England)
log	6 eggshells full of liquid (ancient Babylon)
nail	8 pounds of butter (England)
perch	24.75 cubic feet of building stone (United States)
rotl	about a pound of weight (old Arabia)
seer	about 15 pounds of rice or other grain (Afghanistan)
shekel	a coin weighing the same as 120 grains of wheat (ancient Israel)
shid	a pile of firewood (England)
to	about 18 liters (Japan)
tum	about 3 centimeters (Sweden)

AND WHAT ABOUT THESE?

Scientists have made up some special words for measurements. For example:

nit	blink
nox	crith
noy	spat

A nit and a nox are measures of light. Noy measures noise, of course. A blink measures time — one-ten-thousandth of a day. A crith is a measure of gases. A spat measures distance in space — a thousand million kilometers.

GOING METRIC

Customary System	Metric System	Metric System	Customary System
1 inch	= 25.4 millimeters	1 millimeter	= .039 inch
1 inch	= 2.54 centimeters	1 centimeter	= .393 inch
1 foot	= .3048 meter	1 meter	= 1.093 yards
1 yard	= .9144 meter	1 meter	= 3.280 feet
1 mile	= 1.609 kilometers	1 kilometer	= .621 mile
1 ounce	= 29.573 milliliters	1 milliliter	= about 1/5 teaspoon
1 pint	= .473 liter	1 liter	= 1.057 quarts
1 quart	= .946 liter	1 gram	= .035 ounce
1 gallon	= 3.785 liters	1 kilogram	= 2.2 pounds
1 ounce	= 28.35 grams		
1 pound	= .453 kilogram		

WHAT IS A HAIRBREADTH ESCAPE?

If a racing car skids on the track but keeps going, you can say the driver had a narrow escape. But if a tire blows out in a race and he almost gets killed, that's a hairbreadth escape.

Where did the saying come from? Long ago a hairbreadth was a real measurement. In the country of Siam people measured small things in hairbreadths. And they had some other curious names for measures. Eight hairbreadths equaled one louse egg. Eight louse eggs equaled one louse. Eight lice equaled a grain of rice, and eight grains of rice equaled one fingerbreadth. The fingerbreadth was almost an inch.

This was a very inconvenient system. Suppose you wanted to figure out how many hairbreadths made a fingerbreadth. You had to multiply $8 \times 8 \times 8 \times 8$. Not an easy problem for anyone who had trouble with arithmetic.

At last, about a hundred years ago, the King of Siam decided his country ought to have a better way of measuring. The system he chose was called the metric system, which French scientists had already invented. All measurements in the metric system can be changed from one to another simply by multiplying or dividing by 10, 100, or 1000. That means just putting zeroes on or taking some away. The measures of length are:

10 millimeters = 1 centimeter
100 centimeters = 1 meter
1,000 meters = 1 kilometer

There are also metric measurements for weights and for containers, for temperature, electric current, and other special things.

Of course, it is easier to multiply by 10 than by 8. So Siamese children had less trouble with arithmetic than before. And it was easier for Siamese businessmen to figure out costs and prices when they were buying and selling.

Almost all countries in the world now use the metric system. Most of the remaining countries intend to use metrics some day, and they are slowly giving up their old ways of measuring. The United States has been one of the slowest. Its system is called the United States Customary System. The one used in England and some other countries is the British Imperial System. In both countries, teachers say that children have to spend about two more years learning arithmetic than children spend in countries that have gone metric.

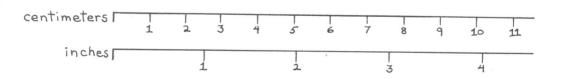

WHY DIDN'T THE UNITED STATES ADOPT THE METRIC SYSTEM WHEN OTHER COUNTRIES DID?

Frenchmen invented the metric system about two hundred years ago, and soon the French government adopted it. From then on everybody in France was supposed to measure things metrically. At first there was a lot of grumbling about the change. But after a while most people got used to the new measures and liked them.

Smart men in other governments saw that the French system was good. Their old ways of measuring were a mishmash of rules and customs hundreds or thousands of years old. So why not change to the new and simple French way?

Some leaders in the United States wanted to go metric. Others were afraid. The American Revolution had recently ended, and the country was very poor. Many people were dissatisfied. Perhaps they would refuse to adopt metric measures, and the states might even become disunited. Besides, businessmen traded mostly with England, and England had no intention of going metric. Some people even objected to the metric measures because they were not mentioned in the Bible. For all these reasons the United States Congress did not make any law adopting the metric system.

Slowly other countries did adopt the system. Scientists, even in the United States and England, began to use metric measures. After a while laws in those two countries said that both the metric system and the customary systems could be used officially. Finally the English-speaking countries were almost the only old-fashioned ones in the world. They knew they would have to give up their customary systems. But not all at once. They decided that complete change might take many years. They began to put both miles and kilometers on road signs. Races were run in meters as well as yards. Later the old words would gradually disappear as people got used to the new ones.

Maybe, some day, someone will say to you, "Grandma and Grandpa, what was a gallon?"

HOW DO WE KNOW THE LENGTH OF A METER?

A perfect way of measuring — that was what the men who worked out the metric system wanted. No more using the length of a king's foot or some sailor's arm. The new measurement had to be fixed in a way that didn't change. But what could that way be?

The earth seemed to have a fairly permanent shape and size. This meant that the distance from the North Pole to the Equator wasn't likely to change. Also, it could be measured by men using special instruments. And so scientists decided that a meter should be 1 ten-millionth of that distance.

Later, scientists found they needed a more precise way of describing a meter. Some of them met together and agreed to measure it by a beam of a certain kind of light. Now they say it is "1,650,763.73 wave lengths of the orange-red line of the spectrum of Krypton 86."

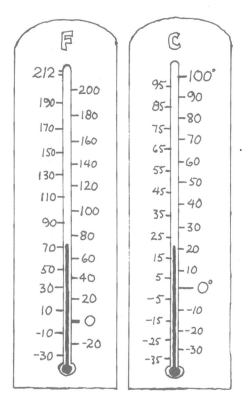

To convert degrees Celsius to degrees Fahrenheit, multiply the Celsius number by 1.8 and add 32.

WHY DO THERMOMETERS HAVE DIFFERENT SCALES?

Most thermometers are sealed glass tubes with liquid inside. The liquid is either mercury, which is silvery, or alcohol, which is usually dyed red. When it is warmed, the liquid expands and rises in the tube. Cooling makes it shrink and go down.

In 1714 an accurate thermometer was invented in Germany by Gabriel Daniel Fahrenheit. He was the first to mark off on a sealed tube an accurate scale for measuring temperature. On this scale the distances between two marks were called degrees. The mark numbered 32 — that is, 32 degrees, also written 32° — shows the temperature at which water freezes. The mark numbered 212 indicates that at 212° water boils at sea level.

Thermometers with Fahrenheit's scale were soon used far and wide. The scale became official in English-speaking countries.

Shortly after this a Swedish scientist, Anders Celsius, worked out another scale. He used zero for the freezing point of water, and one hundred degrees for the boiling point. The Celsius scale is more convenient than the Fahrenheit, because using numbers from 1 to 100 is easier than using numbers from 32 to 212. Most people in the world have adopted the Celsius scale.

Having two scales is a nuisance. Since the Celsius scale is easier to use, the United States Congress has agreed to shift to this scale in the coming years.

HOW DO WE KNOW HOW MANY CALORIES ARE IN A HAMBURGER?

A calorie is a unit — one that we use in measuring heat energy. It is the amount of heat needed to raise the temperature of 1 kilogram of water (about one quart) 1 degree Celsius.

Well, what has that to do with a hamburger? We eat hamburgers, and all other foods, because we constantly use up energy. Energy keeps us alive, and we get it by burning up food. Some foods provide more energy than others. And here's where calories come in. They are the units used in comparing the energy values of different foods.

Scientists find out how many calories are in different foods by burning them up — not in a frying pan on a stove, but in a special instrument. The instrument is called a *bomb calorimeter*. The main part really is a bomb. It is full of pure oxygen and it sits in a tank of water with a thermometer attached. After a tiny piece of hamburger meat is weighed and put into the bomb, an electric spark sets off an explosion. The meat burns up instantly. Heat energy from the meat raises the temperature of the bomb and the water, and that makes the thermometer go up. Since scientists know how much water is in the tank, they can read the thermometer and figure out how many calories were in the tiny piece of hamburger.

But how many calories are there in a nice, big, juicy quarter-pounder? That's about 100 grams of hamburger meat. Scientists have figured out that usually there are about 275 calories in 100 grams of hamburger. And that's not counting the calories in the bun, the ketchup, the relish, the lettuce, the tomato, or the mustard!

Chapter VIII

Round and Square, Thick and Thin

1. *Could a bird sit on a square egg?*
2. *How far away is the horizon?*
3. *Why does the earth look flat when it is really round?*
4. *Why is pi such a strange number?*
5. *Are there any lucky shapes?*
6. *How many of these shapes can you name?*
7. *Why are dice always cube-shaped?*
8. *What is a square deal?*
9. *Can a dog draw a circle?*
10. *Why do starfish have five arms?*
11. *Why are there right angles but no left angles?*
12. *Do stars really have five points?*
13. *Why are bananas curved?*

COULD A BIRD SIT ON A SQUARE EGG?

Probably not long enough to hatch a chick. The bird would last, but the egg would not. A square egg would soon break under the bird's weight.

Some shapes are stronger than others. An egg shape is stronger than a cube. Because their shape is strong, eggs can have thin shells and still not be crushed when a bird sits on them.

A curved shape called an arch is strong, too. The bones in your foot form an arch. Your feet are small compared to the rest of you, but they have to support your whole weight. It is important for your feet not to be too big. Very big feet would get in your way when you wanted to run or jump. The arch shape helps make your feet very strong for their size.

Arches and other curves are used to make strong buildings. A dome is a roof that is like one end of an egg — curved in all directions. It is a very strong shape. Whole buildings can be made in the shape of a dome. They can be built with less material than square buildings that enclose the same amount of space.

Egg shapes, arches and domes all help save space and materials. From little eggs to big buildings, things are made strong by curves and arches.

HOW FAR AWAY IS THE HORIZON?

Suppose you are at the seashore, looking out across the ocean. The line where the ocean and the sky seem to meet is called the horizon. The line of the horizon is as far as you can see. Beyond it things are hidden by the curve of the earth. From the shore you sometimes see a ship's smokestack, but not the rest of the ship. It is below the horizon.

The distance from you to the horizon depends on how high up you are. If you are four feet tall, and you are standing on the beach, the horizon is a little less than two miles away. From the top of a 100-foot lighthouse, the horizon is about 12 miles away.

If you were an astronaut standing on the moon, how far away would the moon's horizon be? It would be closer than the horizon on earth, because the moon is smaller than the earth. Its surface is more steeply curved and so your line of sight stops at a much shorter distance. If a midget astronaut four feet tall was standing on a flat spot on the moon, the horizon would be only about a mile and a quarter away.

WHY DOES THE EARTH LOOK FLAT WHEN IT IS REALLY ROUND?

Compare a basketball and an orange. They are different sizes, but they are the same shape. Both are spheres. But the curve of the basketball is not the same as the curve of the orange. Suppose you cut a round piece the size of a half-dollar from the orange, and imagine you cut a similar piece from the basketball. The piece from the orange will curve noticeably. But the piece from the basketball would curve more gently. It would seem much more flat. A piece cut from a larger sphere, such as a beach ball, would seem even more flat.

Our earth is a sphere, too, but it is much bigger than any ball. The earth is so big that the curve of its surface is very gradual. The part that we can see is only a tiny bit of its surface. Our eyes cannot even detect its gentle curve. And so it seems to be flat.

The earth does look round to astronauts when they go out into space. The farther they go the smaller it looks, so they see more and more of its curvature at one time. In space, they can see the earth's real shape.

WHY IS *PI* SUCH A STRANGE NUMBER?

Suppose you have an automobile tire that measures 1 meter across. How big around is the tire? You can find out by measuring with a tape measure. You can also multiply the distance across — the diameter — by a strange number called *pi*. For practical purposes, pi is about 22/7. Now, if you multiply the diameter, 1 meter, by 22/7, you find that your tire is about 3-1/7 meters around.

But how much is pi exactly? No one really knows! Computer scientists have now computed pi to over one million decimal places. It starts out looking like this:
3.14159265358979323846264338327950288419

Hundreds of years ago mathematicians weren't sure what sort of number pi was. Many of them tried to prove that it was a fraction. When any fraction is written as a decimal number, the same digits always appear over and over again in a special pattern. For example, the fraction 22/7, which is a little bigger than pi, looks like this when it is written as a decimal:
3.142857142857142857142857142857142857

If pi were a fraction, there would be a repeating pattern to its digits. But no one could ever find a repeating pattern in pi. Finally, in 1761, a Swiss mathematician named Johann Heinrich Lambert settled the matter once and for all. He proved that pi could not possibly be a fraction.

Mathematicians are still calculating more places for pi because they want to see if the digits are arranged in any special way. For instance, does 3 show up more often than 0? There is a total of seven 3's in the first 40 digits of pi, while 0 shows up only once. So you might think pi has more 3's in it than 0's. But this isn't so. If you go out a little farther, the number of 3's and the number of 0's seem to even out. In fact, no one has ever found any rhyme or reason to the arrangement of the digits.

Everyone who has known about pi has been fascinated by this remarkable number. *Almost* everyone, that is. Many people in the state of Indiana in 1897 couldn't have cared less about pi. They thought the decimal part was just a nuisance. So they tried to pass a law that would have made the value of pi in their state equal to 3 — exactly!

"Forget .14159265 and the rest of it," they said. "3 is good enough for us." The law almost passed, but in the end it was defeated.

What if the law had passed? With pi exactly equal to 3, automobile tires on Indiana cars might look like this:

ARE THERE ANY LUCKY SHAPES?

Once there was a great teacher who thought that mathematics was wonderful — even magical. His name was Pythagoras, and his students were called Pythagoreans.

The Pythagoreans formed a kind of secret math club. Only men could belong, but women were allowed to listen when the teacher gave lectures. Members of the club believed that a shape called a *pentagram* was especially beautiful. It became the club's symbol for health — and for good luck.

Once a Pythagorean got very sick when he was away from home. Kind people tried to help him, although he had no money to pay them. At last he said, "After I die, paint the pentagram on the outside wall of your house." They did. Many years later another Pythagorean passed the house and saw the symbol. "Why is that here?" he asked.

The people told him the story, and he gave them a large reward.

The Pythagoreans also loved three special solid shapes and thought these, too, brought good fortune. One of the shapes, called the dodecahedron, has twelve sides, all equal — and all of them are pentagons!

How to make a pentagon with a strip of paper:

1. Tie an ordinary "overhand knot" in the strip.

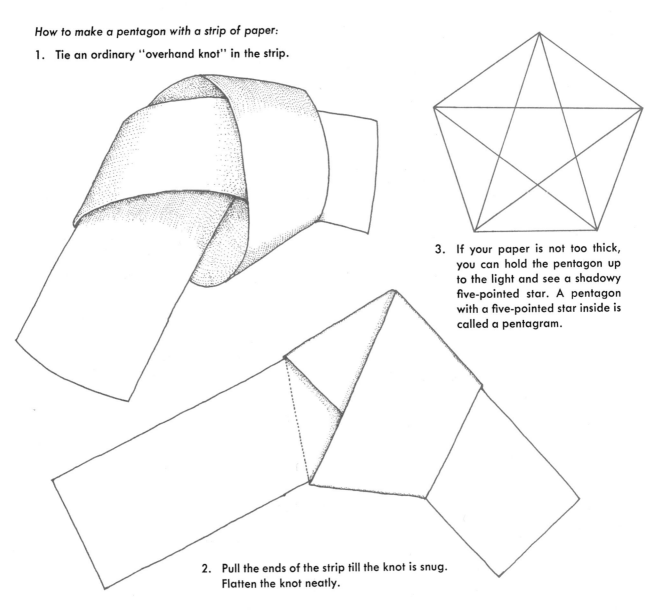

2. Pull the ends of the strip till the knot is snug. Flatten the knot neatly.

3. If your paper is not too thick, you can hold the pentagon up to the light and see a shadowy five-pointed star. A pentagon with a five-pointed star inside is called a pentagram.

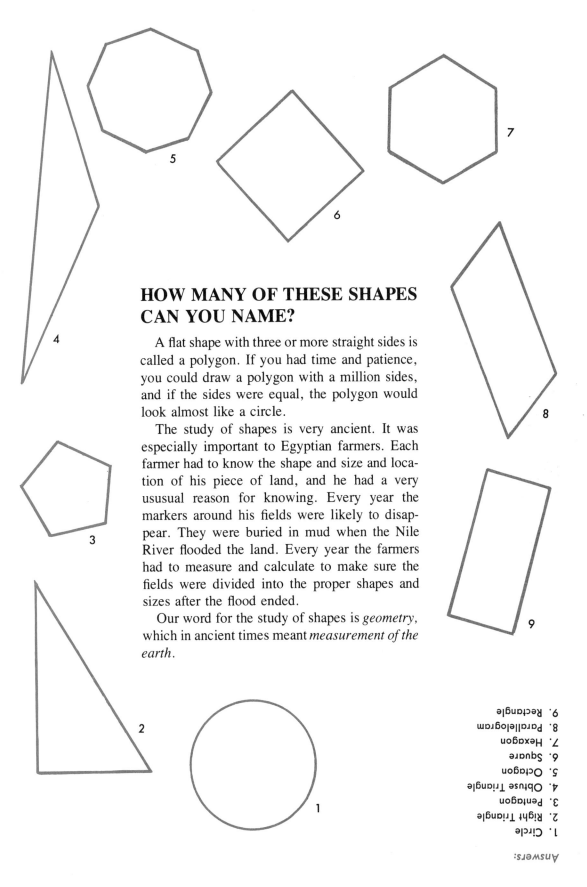

HOW MANY OF THESE SHAPES CAN YOU NAME?

A flat shape with three or more straight sides is called a polygon. If you had time and patience, you could draw a polygon with a million sides, and if the sides were equal, the polygon would look almost like a circle.

The study of shapes is very ancient. It was especially important to Egyptian farmers. Each farmer had to know the shape and size and location of his piece of land, and he had a very ususual reason for knowing. Every year the markers around his fields were likely to disappear. They were buried in mud when the Nile River flooded the land. Every year the farmers had to measure and calculate to make sure the fields were divided into the proper shapes and sizes after the flood ended.

Our word for the study of shapes is *geometry*, which in ancient times meant *measurement of the earth*.

Answers:
1. Circle
2. Right Triangle
3. Pentagon
4. Obtuse Triangle
5. Octagon
6. Square
7. Hexagon
8. Parallelogram
9. Rectangle

85

Pieces used in games of chance:

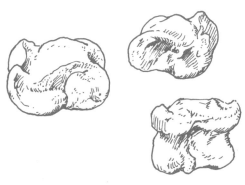

1. Animal knuckle bones were used like dice in ancient times.

2. African carved nut shells.

3. In Egypt, in 1795 B.C., these little carved ivory monkeys were used like dice.

WHY ARE DICE ALWAYS CUBE-SHAPED?

For thousands of years, people have played games with dice, sometimes called number cubes. Each cube must be made as nearly perfect as possible, with all six sides exactly the same. If the shape is imperfect, one side will turn up more often than another, and so the game cannot be played fairly.

The same kind of game could be played with dice made in other shapes, so long as the numbered sides were all exactly the same. You could even put numbers on pencils that have six flat sides and roll the pencils instead of throwing them. But cube-shaped dice are more comfortable to hold and shake in your hand. Another reason for the shape is that it is easier to make a nearly perfect six-sided object than to make one with eight or twelve or twenty perfect sides.

A dishonest player will sometimes load the dice. That means he adds a little extra weight to one side of each die. The heavier sides will land on the table more often than the lighter sides. Of course, the numbers on the lighter sides will then show up more often. This helps the cheater to win.

WHAT IS A SQUARE DEAL?

Ever since ancient times the square has been a symbol of honesty. All its sides are equal, and that made it stand for perfect fairness. Today, if we buy a car for a price we think is fair, and if it runs properly, we are likely to say that the salesman gave us a square deal.

CAN A DOG DRAW A CIRCLE?

All the points on the edge of a circle are exactly the same distance from the center. A dog doesn't have to know this in order to run in a circular path. And you don't have to train him for years, either. Tie one end of a rope to his collar and the other end to a post with a lot of room around it. Then get the dog to run around the post. If the rope stays tight, the dog's feet will make a circular path.

WHY DO STARFISH HAVE FIVE ARMS?

A starfish lives in the sea, but it is not really a fish. A fish has many long, thin bones. But the skeleton of a starfish is made of rather flat separate plates. There is one plate for each arm. In an adult starfish these plates are held together by muscles and a tough outside skin. But a very young starfish has more delicate skin. Its muscles are weaker. And that is where a five-part shape comes in handy. It helps to protect the young starfish. To see why, try thinking of something entirely different. Think about a bunch of bread rolls baked in different shapes.

Here are four rolls arranged in a square.

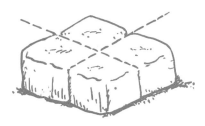

Here are six rolls.

Here are five rolls.

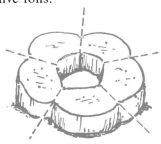

Now suppose you try to break each batch apart along the dotted lines.

In the arrangement of four and six, there are creases opposite one another. This makes it easy to break the batches in half.

But the arrangement of five is harder to break. You have a solid roll opposite any of the creases. So you will have to tear that solid roll apart in order to break the batch in two.

A scientist who has studied starfish thinks that this kind of five-part arrangement gives them extra protection when they are young and delicate. Suppose a little starfish with a five-plate skeleton is struck or pushed. It does not break in two as easily as a four-plate animal would.

Most kinds of starfish that live in the sea today have five-plate skeletons and five arms. One kind has ten arms and another has twenty, but these are multiples of five. Only a few have six. Long ago there were quite a lot of starfish that did not have plates arranged in fives. Scientists have found their fossil remains in layers of rock. But those starfish have died out. Apparently the five-armed kind, with their stronger skeletons, had the best chance of surviving.

WHY ARE THERE RIGHT ANGLES BUT NO LEFT ANGLES?

When you talk about your right hand, right means the opposite of left. But it has a different meaning when you say "The answer is right." In that case right is not the opposite of left. And the same is true of the right in a right angle. That's why there are no left angles.

Long ago, one of the meanings for right was straight up and down. If the walls of a house were straight up and down, they were right. And they made a right angle with the floor.

A right angle came to be any angle with sides that were perpendicular to each other. If one side was perfectly flat, like a floor, the other would point straight up or down. A right angle is a ninety-degree angle. If you tip it sideways, it is still a right angle, even though its sides no longer point up and down.

DO STARS REALLY HAVE FIVE POINTS?

Stars only seem to have points. A star's twinkling makes us think we are seeing points of light. Because a five-pointed shape is one that many people like very much, artists often draw stars in that shape.

Many things in nature do have five points or five sides or five parts. Flowers with five petals are common. The human hand has five fingers. So does a monkey's hand. A cat's paw has five toes. And an earthworm has five pairs of hearts!

WHY ARE BANANAS CURVED?

When a tiny banana begins to grow, one side grows a little faster than the other. The fast growing side of the banana becomes longer than the slow side. It is the difference between its two sides that makes the banana curve. The longer side of the banana curves around the shorter side.

Both sides taste the same, but there is a little more to eat on the long side.

Snail shells are also curved, because one side grows faster than the other.

Chapter IX

What Is It Worth?

1. *What does unit price mean?*
2. *Is anything cheaper by the dozen?*
3. *Why is 13 a baker's dozen?*
4. *What do numbers on a milk carton mean?*
5. *How do you read an electric meter?*
6. *What is a sales tax?*
7. *What is pin money?*
8. *What is a budget?*
9. *What is double-digit inflation?*
10. *Why does a phone bill have holes in it?*
11. *How does a credit card work?*

WHAT DOES UNIT PRICE MEAN?

A unit is one complete thing. A pound is a unit of weight. An ounce and a liter are units, too. Unit price tells you the cost of one pound of butter or one ounce of perfume or one liter of gasoline.

Perhaps that doesn't sound important. It wouldn't be important except for one strange thing about human beings. Our eyes often fool us. We can look at two different bottles of cooking oil, for example, and the tall one will seem to hold more than the short one. But does it? The tall bottle actually holds 10 ounces. But the short one holds 12.

Now suppose that the two bottles cost the same — $1.20 each. How can you tell which one to buy? You can tell by finding out the unit price of each ounce.

Take the tall bottle first. Its 10 ounces cost $1.20. So one ounce costs $1.20 ÷ 10, which is 12 cents. Each ounce in the short bottle costs $1.20 ÷ 12, or 10 cents. The unit price for the tall bottle is 12 cents and for the short bottle 10 cents. So you can save money by buying the short bottle.

Usually it is not so easy to figure out the unit price. A bottle that costs $1.53 may hold 21½ ounces. It would take most people a long time to figure out whether that bottle is cheaper' than another bottle of a different size. So some stores help you solve the problem. They print the unit price on the shelves above or beneath packages. That way customers can tell whether one kind of oil or coffee or candy is more expensive than another.

Wouldn't it be better just to make packages regular sizes and avoid all the bother? It certainly would be better for customers, but product manufacturerers believe they would make less money. They can be sure that many buyers never look at unit prices, so they continue making packages that fool the eye and cost more.

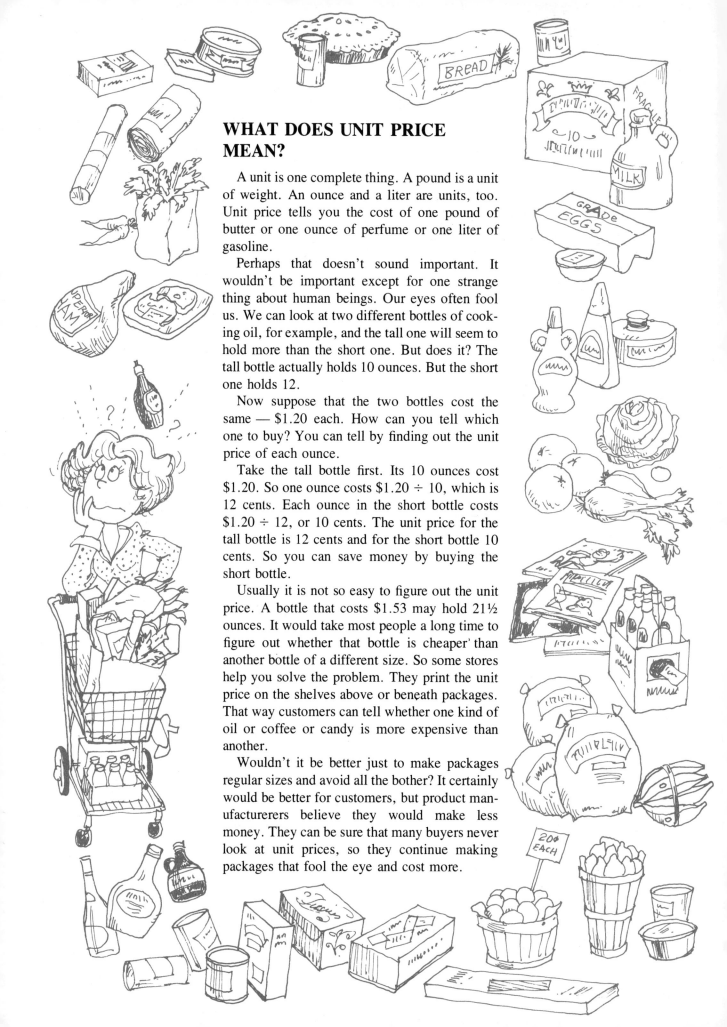

IS ANYTHING CHEAPER BY THE DOZEN?

There was once a storekeeper who liked to play jokes on customers. If you asked him how much his cookies cost, he might say, "Five cents apiece; sixty-five cents a dozen." Unless you were paying attention, you probably would think he was offering you a bargain. So you would buy a dozen. Yet, if you paid five cents each for twelve cookies, the price would be sixty cents a dozen, not sixty-five cents. If you didn't stop to figure that out, the storekeeper wouldn't tell you.

Some things really do cost less if you buy several or a dozen instead of one. The storekeeper can charge less because he puts everything in one bag instead of twelve bags. He spends less time selling a dozen all at once. And he makes money faster by selling things faster.

In countries where almost everything is made by machines in factories, people are used to the idea of "cheaper by the dozen." But in places where craftsmen make things by hand, that is not always so. Once a woman who lived in a big city went to a small village where a carpenter made chairs by hand. She saw one she liked and asked the price. The man told her, and she said, "Please make me five more just like it."

"In that case," the man said, "the price will be much higher. It is very boring to make things exactly alike. So I will have to charge you more." The woman agreed. More does not *have* to mean cheaper.

WHY IS 13 A BAKER'S DOZEN?

Bread is such an important food that many countries have laws to make sure people can buy good bread at a fair price.

In the old days a law in England said that if a baker sold rolls by the dozen, the 12 rolls had to weigh a certain amount. Wise bakers did not want to be accused of cheating their customers. So they often threw in an extra roll with every dozen, just to make sure the buyer got the weight she was paying for. After a while, 13 of anything was called "a baker's dozen."

WHAT DO NUMBERS ON A MILK CARTON MEAN?

If you should see Ju476 on a milk carton in a store, don't buy it! The letters and numbers stand for June 4, 1976. The milk company stamped them on the carton to show that the milk was supposed to stay sweet until that date. After that the storekeeper was not supposed to sell it. Of course, you are not likely to find milk that has been for sale for years. Most stores are careful not to sell spoiled milk.

Dates printed on packages of many kinds are supposed to help customers get food that is reasonably fresh. But some of the numbers are hard to understand. They are written in a kind of code which only the storekeeper and the manufacturer recognize. The number 006467 can mean June 4, 1976 in secret code. The zeroes help to fool you. The first 6 stands for the sixth month; 4 is the day of the month; the year '76 is written backward (67) to make the code more difficult. So, if a package is really older than it should be, you may not find it out, unless you are a good detective — or unless the food doesn't taste good.

HOW DO YOU READ AN ELECTRIC METER?

The dials on your electric meter look a little bit like four clock faces. But each face has only one hand, and the numbers on the faces go from 0 through 9 instead of from 1 through 12.

Take a look at the first face. On that face and on the third face the numbers go counter-clockwise. The hands on the dials also move counterclockwise.

If you want to read your electric meter, just read the numbers on the dials.

The reading on this meter is 6875. When a hand points between two numbers, the correct number is always the smaller number.

An electric meter measures a complicated quantity called kilowatt-hours. If you turn on ten 100-watt light bulbs for one hour, you will use 1 kilowatt-hour of electricity.

How many kilowatt-hours are used at your house in a month? Look at your meter, and write down the number shown on the dials. Let's say it's 6,000. A month later, read the meter again. Let's say this time it's 6,850. The difference between the two readings is 850. That is the number of kilowatt-hours your family used during the month.

Very few of those 850 kilowatt-hours went for light bulbs. Most of them probably went to run things like a refrigerator or an air conditioner. Some people have electric hot-water heaters, electric stoves, and electric heating systems. Each of these uses a lot of electricity, too.

WHAT IS A SALES TAX?

You see a nice pen and decide to buy it. The price is marked $1.00. But in places where there is a sales tax, the pen will cost more than that. If the tax is 5%, you must pay an extra five cents, bringing the cost to $1.05. If you are buying other things, you will have to pay the 5% tax on them, too.

Who will get the tax money? In the United States, the state governments usually get it. In some states, the sales tax is 3%; in others it is 4 or 5%. Many cities also have a sales tax which must be paid in addition to the state tax.

A sales tax was used in Spain at the time of Columbus. Eventually it was dropped because it hurt business. It was not used again until this century when money was needed to help pay the cost of World War I. At that time Canada, France, Germany and Italy began to collect a sales tax. In the United States, the tax was started to help pay relief money to people who lost their jobs during the depression in the 1930's. Now sales taxes are found nearly everywhere, and they are used to raise money for many different purposes.

WHAT IS PIN MONEY?

When your grandmother was a girl, she probably got presents of a few coins on her birthdays. She could spend the coins for little things she wanted, and they were called pin money, though she may not have known why.

The first women who got pin money really did spend it for pins. That was several hundred years ago in England. In those days metal pins were quite a new invention. You could not buy them every day and they were very expensive. In fact, they were so scarce that storekeepers were allowed to sell them only on January 1 and 2. Since women never had any money of their own, they had to ask a man in the family for enough to buy pins on New Year's Day.

After a while women could buy as many pins as they needed any day in the year. Still, they did not give up the habit of asking for pin money which they spent for other things as well. Finally, the words came to mean any small amount a woman could use as she pleased.

A man with a big belly holds a bulging bag of money that he wants to budget. The words *belly, bulge,* and *budget* all have the same ancestor — the Roman word *bulga,* which means *bag.*

WHAT IS A BUDGET?

In some places in the world women are not supposed to have jobs and earn money. If they want to buy food or clothes, they have to ask their husbands or some other man in the family. That used to be the rule in many places.

In ancient Rome, one of the chores women had to do was to pay the family's bills. But men often failed to give them enough money at the right time. So Roman women worked out a system. If a wife had to pay for six different things, she put labels on six little leather bags. Then she divided up whatever money her husband gave her. She put a little in one sack, more in another, depending on which bills were likely to be big and which ones small. A careful wife could manage to have enough money in a bag to pay a bill on time.

The word for bag in the Latin language, which Romans spoke, was *bulga*. As time passed, people in France borrowed the word but pronounced and spelled it differently — *bougette*. People in England learned the word from the French. They also borrowed the habit of figuring out their expenses, then saving the exact amount of money needed to pay each bill. This careful plan was finally called a budget.

Most people now keep their money in a bank, not a bag. Governments also keep money in banks, and governments also have budgets. But the Roman woman's *bulga* was the ancestor of the word "budget" and of the budgeting idea.

WHAT IS DOUBLE-DIGIT INFLATION?

We use the word *digit* when we talk about numbers. For example, 357 is a three-digit number, and 12 is a two-digit number. Double-digit means two-digit, but it sounds trickier.

Inflation is tricky, too. Suppose you own an ice-cream store and you charge 40 cents for a chocolate cone. One day you decide you can make more money if you charge 50 cents. Soon all the ice-cream stores decide to raise their prices, too. So do the restaurants and movie theaters. That is inflation.

Perhaps the difference between 40 cents and 50 cents does not seem very great. But ten cents is actually one-fourth of 40 cents. Another way to say one-fourth is 25%. When you add 25% to the price of an ice-cream cone, that is double-digit inflation because there are two digits in 25%.

When everything in a whole country costs more and more, inflation is a bad problem. It often starts because some people see a chance to make more money. Soon the men and women who work need more wages in order to pay the higher prices, and the problem gets worse and worse. Even the experts can't agree on the best way to stop inflation.

WHY DOES A PHONE BILL HAVE HOLES IN IT?

A telephone bill in the United States and some other countries often comes in two parts. One part lists the cost of the phone calls. The other part is a card covered with strange rectangular holes. It is called a *punched card,* and the holes are arranged according to a special code. The code tells who you are, what your phone number is, and how much you owe the phone company.

When you pay your bill, a clerk puts the punched card into a machine called a card-reader. The card-reader sends out electrical signals. These signals are stopped wherever there are no holes in the card. But wherever there are holes, signals can pass right through. Every special pattern of holes becomes a special pattern of electrical signals. Then a computer decodes the signals and makes a record of the amount you have paid.

All this can happen very quickly. Some card-readers can "read" a thousand cards in a single minute!

HOW DOES A CREDIT CARD WORK?

Suppose you go to a sports shop and buy a pair of ice skates. Then you buy some books at a book store. And then you buy a new tire for your bicycle at a bike shop. If you are old enough to use a credit card, you can make each of these purchases without using any cash. All you have to do is sign your name to a little slip of paper in each store. The paper lists all the items that you bought in that store. The clerk gives you a copy of the paper, sends a copy to the credit card company, and keeps one copy for the store.

At the end of the month the credit card company's computer goes to work. It takes its record of your monthly purchases and adds up the money you owe. Then it automatically prints a bill which the company mails to you. When your bill comes, you write just one check to pay for the skates, the bicycle tire, the books, and any other credit card purchases you may have made. Finally, the credit card company cashes your check and sends money to the various stores to pay for the things you bought.

How does a credit card company make any money from all this? When you buy something with your credit card, not all of the money you spend goes to the store that made the sale. About 3% of your payment goes to the credit card company. If the books you bought cost 10 dollars, the company sends the bookstore only $9.70, keeping 30 cents for itself.

In many places your credit card bill is actually a little more than the cost of the things you purchased. This extra charge is called a finance charge. The finance charge adds a little more money to the profit the credit card company makes. And it also makes credit-card purchases a little more expensive than purchases you pay for with cash.

Chapter X

On the Road

1. *How can you figure miles per gallon?*
2. *Why can you go farther on a gallon of gas in Canada than in the United States?*
3. *How far can a car travel on a gallon of gas?*
4. *Is a speedometer the same as an odometer?*
5. *Why are there both letters and numbers on some license plates?*
6. *How do you read the mileage table on a road map?*
7. *What does a "measured mile" mean?*
8. *When was a mile not a mile?*

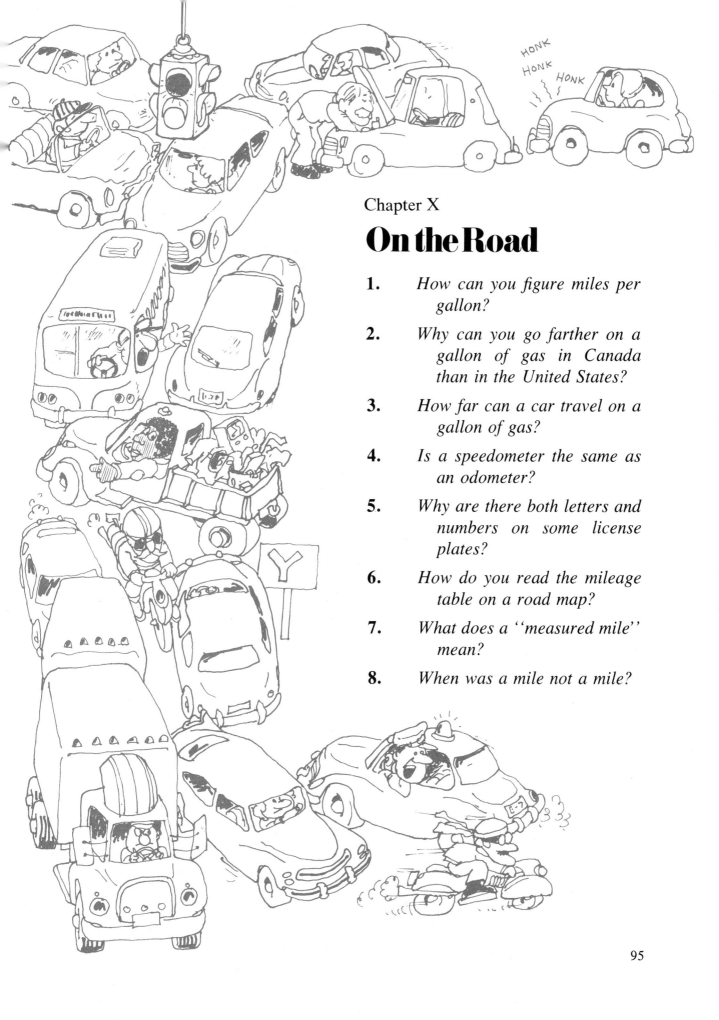

HOW CAN YOU FIGURE MILES PER GALLON?

"Fill 'er up!"

If you want to know how many miles per gallon your car gets, start figuring at the service station. While the gas tank is being filled, look at the number on the odometer. If it says 20,000, the car has traveled a total of 20,000 miles. Write this number down and keep it.

The next time you stop to buy gas, be sure to find out how many gallons it takes to fill the car's tank. If you need 9 gallons, then you must have used 9 gallons since your last fill-up.

Look at your odometer again. Suppose it shows 20,270 miles. Now you can do your arithmetic.

```
  20,270 miles second fill-up
 -20,000 miles first fill-up
     270 miles traveled between fill-ups
```

You have gone 270 miles on 9 gallons. How many *m*iles did you go *p*er *g*allon (m.p.g.)? To find out, divide 270 miles by 9 gallons. The answer is 30 miles per gallon of gas. In case you forget which number to divide by, use the m.b.g. rule: Divide the number of *m*iles *b*y the number of *g*allons.

$$M \div G = MPG$$

WHY CAN YOU GO FARTHER ON A GALLON OF GAS IN CANADA THAN IN THE UNITED STATES?

Is it because the miles are shorter or because your car runs mysteriously better in Canada? No! It's because the gallons are actually bigger. In Canada a gallon is 1.2 times as big as a gallon in the United States, and it is called an imperial gallon. This means that if your car goes 30 miles on every United States gallon, it will go 30 × 1.2, or 36 miles on every imperial gallon. Perhaps you think that this is a big bargain. But service station owners in Canada know that their gallons are bigger than U.S. gallons, and they charge more for them!

HOW FAR CAN A CAR TRAVEL ON A GALLON OF GAS?

Gasoline is expensive, and the price keeps going up. When prices are high, many drivers pay careful attention to the distance they can drive on a gallon of gas.

The size of a car affects the gas mileage. Big cars are very heavy — some weigh over two and a half tons. They are very comfortable and very powerful, but they use a great deal of fuel. The biggest cars often go less than 10 miles on a gallon of gas.

Little cars do better because they weigh much less than big cars. Small cars may weigh less than a ton. Because they are so light, they can have much smaller engines. Some little cars with small engines can go over 40 miles per gallon.

Is this the best gas mileage possible for the average car? Most people don't think so. And some scientists have come up with clever ways to help drivers go farther on a gallon of fuel. In one plan, alcohol is added to gasoline. In another, water is mixed thoroughly with gasoline in a special way. Water in gasoline usually makes a car run badly or stop running altogether. But with this special mixture, cars run much more efficiently. They go farther on each gallon, and they also cause less pollution because the water helps the gasoline burn better. Scientists hope these schemes will greatly increase gas mileage.

IS A SPEEDOMETER THE SAME AS AN ODOMETER?

A speedometer tells how fast your car is going. An odometer tells how far your car has gone. The odometer is usually located near the speedometer. There is a good reason for this.

The odometer is connected by a long cable to your car's transmission. As the car moves, the cable rotates. And the whirling cable makes the numbers on the odometer dial change.

The number farthest to the right changes after every one-tenth mile you drive. The other five numbers tell you how many miles you have traveled. When you have gone 100,000 miles, six zeroes appear on the dial and the odometer starts counting over again.

The same long cable that runs the odometer is also connected to a strong circular magnet inside the speedometer. The magnet rotates, and as it turns, it makes the speedometer needle move. Some speedometers show speeds as high as 120 miles an hour. But no sensible driver goes that fast on the highway.

WHY ARE THERE BOTH LETTERS AND NUMBERS ON SOME LICENSE PLATES?

Suppose an automobile license plate has seven numbers printed on it. Using combinations of seven numbers, it is possible to make ten million different plates. The first would be 0000000 and the last 9999999. But some countries and some big states, such as California, have more than ten million cars. If California plates were made only with numbers, there would not be enough to go around. So both numbers and letters are used. This means that many more plates can be made. In fact, with both numbers and letters, about 70 billion combinations are possible.

There is another important reason for using letters on license plates. In some states, and in Europe, the letters tell what city or county the car comes from. HA on a German license plate tells you that the car is from Hannover.

In some places people who pay an extra fee can make up their own license plates. These are called "vanity plates."

HOW DO YOU READ THE MILEAGE TABLE ON A ROAD MAP?

When you plan an automobile trip from one city to another, it's a good idea to know how far you will have to drive. Sometimes you can figure out this distance with the help of a road map mileage table. These tables are big lists of numbers. Each number stands for the driving distance between two cities.

	MEMPHIS	MIAMI	NEW YORK	PITTSBURGH	ST. LOUIS
MEMPHIS		1017	1142	786	294
MIAMI	1017		1327	1237	1222
NEW YORK	1142	1327		363	961
PITTSBURGH	786	1237	363		599
ST. LOUIS	294	1222	961	599	

Notice the number 1237. It is in both the Pittsburgh row and Miami column. You can also find it in the Miami row and the Pittsburgh column. That's because the distance between Miami and Pittsburgh is the same in both directions, and that distance is 1237 miles.

Now suppose you want to find the driving distance from New York to St. Louis. First run your finger along the New York row until you come to the St. Louis column. The number in that column — 961 — is the distance between New York and St. Louis. The other numbers in the New York row stand for the distances between New York and the other cities in the chart. The blank space in the New York row also comes in the New York column, and it means 0. What else could the distance from a city to itself be?

WHAT DOES A "MEASURED MILE" MEAN?

Suppose the odometer tells you your car has gone 9,000 miles. Since automobile odometers aren't always accurate, how can you tell how far your car has really gone? You can find out by using a "measured mile."

A measured mile is a one-mile distance on the highway that workmen have measured very carefully.

Just as you pass the "Begin Measured Mile" sign, look at your odometer. Suppose it shows 9,000.0 miles. Watch for the "End Measured Mile" sign. When it goes by, check the odometer again. Suppose it shows 9,000.9. This means that for every mile your car travels, your odometer only records .9 miles. So your car has actually gone farther than the mileage you see on the odometer. Now if you divide .9 into 9,000.9 miles, you will find the true distance your car has traveled: 10,001 miles.

But 10,001 miles may not be a completely accurate figure, either. For one thing, it's hard to read an odometer exactly. Tire size makes a difference, too. An odometer works by counting the number of times the wheels go around. With big tires or snow tires, your car goes a little farther during each wheel revolution than it does with small tires.

WHEN WAS A MILE NOT A MILE?

When you walk a mile you go 5,280 feet. But a mile wasn't always that long. Two thousand years ago Roman soldiers were trained to march with precise steps. Two steps were called a pace, and 1,000 paces made a mile. The word "mile" comes from the Romans' word *milia* — thousand. The length of a soldier's pace was 5 feet, and so a mile was 5,000 feet.

Wherever the Romans went, they built and measured roads. Even after the Roman armies left what is now England, the English people kept on using Roman roads and measurements. They also used measures that weren't borrowed from the Romans. One was the yard — the length of the sash a king wore around his waist. Another measure was the furlong — 220 yards. Furlongs, however, caused trouble. You couldn't divide a Roman mile into an even number of English furlongs. Finally, Queen Elizabeth I, about 400 years ago, decided to change the length of the mile. She said it should be exactly 8 furlongs — 5,280 feet. And a mile is still 5,280 feet. But that mile, like older ones, will become ancient history when the metric system is used everywhere.

Chapter XI

Time and Space

1. *Why does the calendar have twelve months?*
2. *How do we know the length of the year?*
3. *Why are there seven days in a week?*
4. *Why isn't Easter on the same date every year?*
5. *Why do Jewish holidays come on different dates from year to year?*
6. *Why are there 24 hours in a day?*
7. *Why are there 60 minutes in an hour?*
8. *What is a great-circle route?*
9. *What is jet lag?*
10. *How can you lose a day?*
11. *What is a light year?*
12. *What do B.C., A.D., and B.P. mean?*

WHY DOES THE CALENDAR HAVE TWELVE MONTHS?

Long ago, when people were living in caves, they noticed that the moon changed regularly. From a thin sliver of light in the sky, it grew into a full round disc. Then it dwindled away and disappeared for a day or two until a thin new moon appeared.

After a while moon-watchers began to keep track of the changes. Writing had not yet been invented, but it wasn't necessary to put down words. A person could take a bone and a sharp stone and cut a nick in the bone every night between one new moon and the next. Today's scientists know this was done in Europe about 13,000 years ago. They have found two ancient bits of eagle bone marked with cuts, and the cuts correspond to moon changes.

In many places, ancient sky-watchers noticed that there were about twelve new moons between the beginning of one winter and the beginning of the next. And so they began keeping track of the year by means of the moon.

After the invention of writing, sky-watchers began to prepare calendars. They used the period from one new moon to the next in dividing the year. They knew twelve moons did not exactly equal one year. But they got around this. In Babylonia and in Egypt, calendar makers used twelve moons with extra days thrown in as feast days.

This way of dividing the year is still the basis of our calendar. But instead of calling the twelve divisions of the year moons, we call them months.

HOW DO WE KNOW THE LENGTH OF THE YEAR?

In North America certain Indian farmers were careful sun-watchers. Every day, the sun came up from behind a row of peaks to the east. But not always from behind the same peak. In early winter, they noticed, it rose at the southern end of the mountain range. As spring approached, it rose farther to the north. At last, one summer

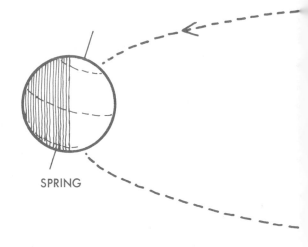
SPRING

day, it seemed to pause. After that, it came up farther and farther south. Finally, in winter, it began its northward journey again.

In this way the Indian farmers kept track of the seasons. Sunrise behind a certain sharp rocky point told them it was time to plant crops. Behind another peak it meant harvest time, year after year.

Farmers in other parts of the world also measured the year by the sun. In some places they watched the shadows at noon. Shadows were longest one day in midwinter, shortest one day in midsummer. Between midsummer one year and midsummer the next year there were about 365 days. Later, more accurate watchers said it was 365¼ days between one midsummer and the next.

In all of this time most people thought the sun traveled around the earth. In Europe they thought so until five hundred years ago. Then astronomers and mathematicians proved that the earth revolves in an orbit around the sun. And the year is actually the time it takes for the earth to complete its orbit.

Since then scientists have developed more and more accurate instruments for observing and measuring. Their measurements show that the earth's journey around the sun takes 365 days, 5 hours, 48 minutes, and 46 seconds. Astronomers call this the solar year.

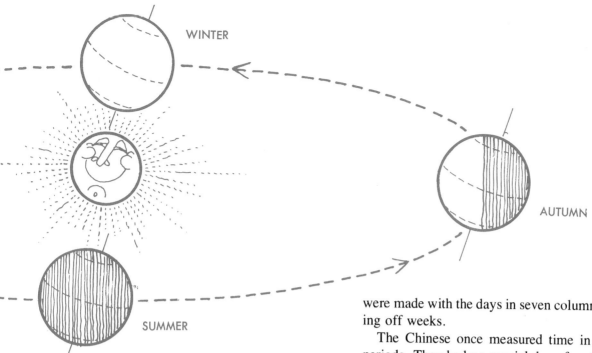

WHY ARE THERE SEVEN DAYS IN A WEEK?

Long ago, in the country of Sri Lanka, priests who followed the Buddhist religion adopted a seven-day rhythm. They got this idea from watching the changes in the moon's appearance. First the moon is small and crescent-shaped. About seven days later it looks like half a disc. Another seven and it is a full disc, then half-full, then a crescent again. Every seventh day in Sri Lanka people rested and sometimes had religious celebrations.

The Hebrew and Christian Bible says that the world was created in six days, and the seventh day was a day of rest. For that reason Jews and Christians long ago arranged their lives according to a seven-day rhythm. Finally calendars were made with the days in seven columns marking off weeks.

The Chinese once measured time in ten-day periods. They had no special day of rest, except for their frequent feast days.

In Africa and South America people followed a rhythm based on the number of days between trips to buy and sell at markets. Their weeks were three or four or five days long.

Nearly two thousand years ago, Christian beliefs began to spread throughout Europe, along with the seven-day cycle. Later, when Europeans settled in North and South America, the seven-day week went with them.

Still later very powerful businesses grew up in Europe and the United States. Business offices were almost always closed on the Sabbath, which was the first day of the week according to the European calendar. Today almost all countries now trade with each other, so most business offices, everywhere, have adopted the seven-day week.

This is how the moon's shape changes from week to week.

WHY ISN'T EASTER ON THE SAME DATE EVERY YEAR?

In spring and fall there comes a time when day and night are the same length. This is called the equinox. The spring equinox has often been a time of celebration. People who lived in cold countries were happy that warm days would make plants come to life. The Christians of Europe rejoiced in springtime for another reason. Spring, they believed, was the season when Jesus came back to life after he was killed. The festival of Easter celebrates this event.

When Easter was first celebrated, religious leaders used a special kind of calendar. Their year had twelve months, but the months had only 29½ days each. That was the number of days between one full moon and the next. According to this calendar there were only 354 days in a year.

At that same time, many people in Europe used a calendar almost like the one commonly used today. It was divided into seven-day weeks, with 365 days in an ordinary year. The months were arranged so that the spring equinox always came on the same date, March 21.

You can see that the moon-calendar with 354 days would get out of step with the ordinary calendar that had 365 days and 366 in leap years. Nevertheless, the Christian priests did not want to set the date for Easter according to the ordinary calendar. So they made a compromise. They decided that Easter should be the first ordinary calendar Sunday after the full moon which appears on or after the spring equinox. Sounds complicated? It is.

Christians now use the ordinary 365-day calendar for almost everything they do. But the date for Easter is still calculated according to the old moon-calendar year. Easter can be as early as March 22, and as late as April 25.

In many parts of the world people celebrate the coming of spring.

WHY DO JEWISH HOLIDAYS COME ON DIFFERENT DATES FROM YEAR TO YEAR?

Jewish religious holidays began in very ancient times. Some of them probably started more than three thousand years ago when the Hebrew people were wandering shepherds. Like many other people who lived in western Asia, the Hebrews measured the year by the moon. Each of their 12 months was just about as long as the time between one full moon and the next.

But 12 moons don't equal an exact year. And so, to make their calendar more accurate, the Hebrew religious leaders decided to add an extra month now and then. Seven out of every 19 years had to have 13 months.

The ancient changeable moon-calendar is called the lunar calendar. The calendar most used today is called the solar calendar because it is based on the number of days it takes the earth to go around the sun. Today Jewish people use the solar calendar in everyday life. But they celebrate religious holidays according to the lunar calendar which has a variable number of months. And so a holiday that always comes on the 14th day of the first lunar month may come on any one of 26 days in March and April, according to the solar calendar.

WHY ARE THERE 24 HOURS IN A DAY?

In all countries where people use modern clocks, a day is divided into 24 hours. Why 24 and not 20 or 100? Nobody really knew the reason for this until scientists began to study certain small pictures on the lids of ancient Egyptian coffins. The pictures, which had a religious meaning, were also star-clocks! They showed how 12 stars, or groups of stars, appeared in the sky, one after another, during the night. This meant that in Egypt more than three thousand years ago, the night was divided into 12 parts. Anyone who knew where to look could tell time by the positions of the stars in the sky.

Time-keeping during the day was different. Egyptians divided the day into ten parts marked off by numbers on a sundial. As the sun slowly moves across the sky, it casts a moving shadow on the numerals on the sundial. This shadow was used to tell the time. Still, there is always a shadowless period before sunrise and another such period at twilight. But the clock-stars are not yet visible at dawn and at twilight. So the Egyptians added these two in-between periods, dawn and dusk, to the ten-part day and the twelve-part night, making a total of 24 parts.

Other people adopted the Egyptians' idea of 24 divisions in a day. And when clocks were invented, they were made to measure off 12 equal hours, twice in each day.

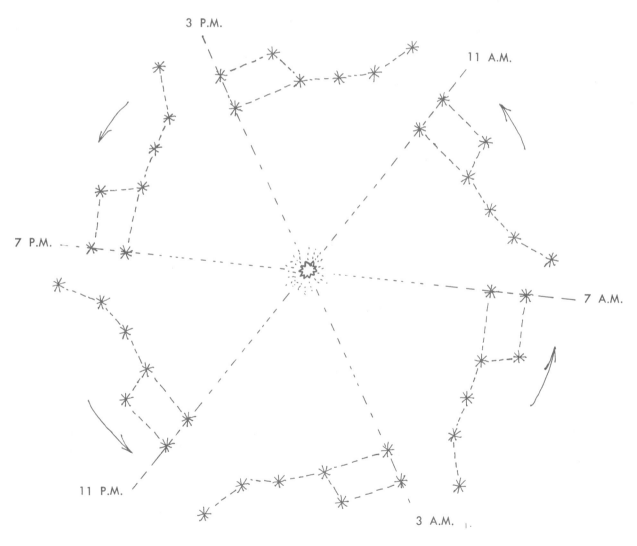

During 24 hours, the Big Dipper seems to make a full turn around the North Star. This is why it can be used as a clock.

103

WHY ARE THERE 60 MINUTES IN AN HOUR?

People invented money before they invented minutes. And the story of minutes begins with the story of money. In very early times some men had only a small amount of silver money, but others had a great deal. Those who were rich could use fairly big weights to measure their silver. A large unit of measurement was called a *mana*. But most people needed small units. They called the small unit a *shekel,* and 60 shekels

A shekel was both a weight and a coin.

made a mana. The number 60 was manageable because it could be divided into many small whole parts — 2, 3, 4, 5, 6, 10, 12, 15, 20, and 30.

The number 60 became so important that it was used for measuring other things besides money. Astronomers decided that each hour of the day should have 60 parts, and each of these should also have 60 parts.

Later on, a scholar whose language was Latin got around to writing about these divisions of the day. When he talked about a "little part of an hour" he called it a *pars minuta*. That got shortened and changed to *minute*. He said the next smaller division of time was the second part — *pars secunda*. And so we have seconds as well as minutes.

WHAT IS A GREAT-CIRCLE ROUTE?

Suppose you are an airplane pilot and you want to fly from San Francisco to Moscow, in the Soviet Union. How would you go?

You might fly from west to east, over Chicago, northern Maine, and London. Or you might fly the great-circle route.

The great-circle route from San Francisco to Moscow goes north instead of east. It goes over Canada and northern Greenland. It passes close to the North Pole. Then it heads south, over Finland, to Moscow. A great circle is any circle around the globe that divides the surface of the earth exactly in half. The equator is a great circle. So is any circle that goes from the North Pole to the South Pole and back around the opposite side to the North Pole.

If you took the west-to-east route between San Francisco and Moscow, you would have to fly 7,500 miles. But the great-circle route is only 5,880 miles long.

There is a great-circle route between any two cities in the world and it is always the shortest route. You can easily find this route if you have a globe and a piece of string. Hold one end of the string down on one of the cities. Then take the loose end, pull it tight and hold it down on the other city. The path that the string traces over the globe is the great-circle route between the two cities.

WHAT IS JET LAG?

When it is daytime on one side of the earth, it is night on the other side. When you board a jet plane in Tokyo, Japan, at 8 o'clock in the morning, clocks in Frankfurt, Germany, say it is midnight. After fourteen hours of flying you land in Frankfurt. If you haven't set your watch ahead, it is still keeping Tokyo time, and it says 10 o'clock (evening time). Then the captain announces the local time. In Frankfurt it is 2 o'clock in the afternoon. You probably have been awake for at least 15 hours already, and you still have a long day ahead.

By 6 o'clock in the evening you will be ready to go to bed, but it will only be time for dinner. And the next morning you may wake up hours before your German friends. You will be very tired for a few days.

This happens because your body follows a kind of built-in time schedule. Usually, sleep ends not long after sunrise and begins a few hours after dark. But on your flight from Tokyo, this schedule is upset. The time when you sleep no longer follows your usual pattern of night and day.

The tired feeling that you get after a long plane ride, when your body is out of step with local time, is called *jet lag.*

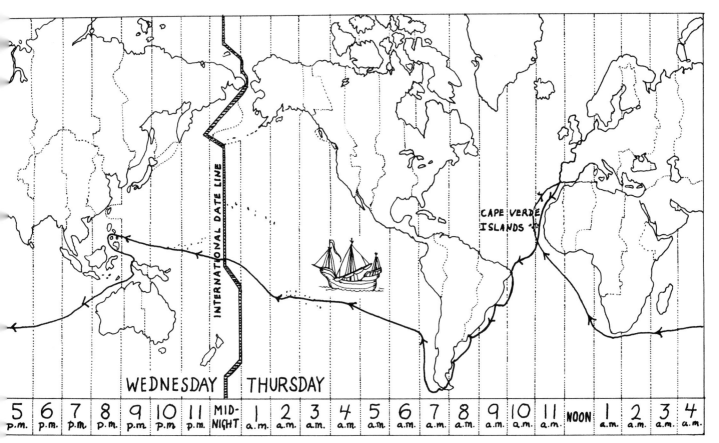

How the date changes at the International Date Line.

Jet lag isn't so bad if you fly only once in a while. But it is a serious problem for people who fly frequently. Because of jet lag, stewardesses and pilots who fly overseas are usually allowed to work only two days a week.

HOW CAN YOU LOSE A DAY?

If you travel around the world from east to west, you will lose a day. The first sailors to do this went with Ferdinand Magellan on his voyage around the world. An Italian named Antonio Pigafetta went along on that dangerous expedition just for the excitement and adventure. Pigafetta kept a daily record during the journey which lasted from 1519 to 1522. Five ships started out, but only one returned home. Toward the end of the voyage his ship stopped at one of the Cape Verde Islands off the coast of Africa, and there Pigafetta was told that the day was Thursday.

"Impossible!" he said. According to his diary, it was Wednesday. He knew he had not skipped a single day. Nevertheless, the people on the island were right. It was Thursday. What happened to the missing day?

The ship had been traveling west. As Pigafetta sailed slowly along, he did not notice an important thing about the time of day. Look at the map and you will see that time changes from place to place. Suppose Pigafetta sailed from Spain one Thursday when the sun was directly overhead. It was noon there. But at that same moment the sun was just coming up farther west in Mexico. There it was only 6 A.M. Still farther west, in Hawaii, it was midnight. And even farther west, it was still Wednesday in Japan!

Pigafetta's ship made many stops, and these time-changes were spread out over three years, so nobody noticed them. But when the ship had gone all the way around the world, the small differences in time between one place and another finally added up to 24 hours. So Pigafetta had been traveling one whole day longer than his records showed.

If Pigafetta could go around the world today, he would find a special place for changing the date in his diary. It is an imaginary line in the Pacific Ocean called the International Date Line. People everywhere have agreed that when it is Wednesday midnight just east of the Line, the calendars and clocks just west of the Line will say midnight, Thursday.

Spiral nebula

WHAT IS A LIGHT YEAR?

A light year doesn't measure time, as you might think. It measures distance. One light year equals about six trillion miles. This is the distance light travels in a year. And so it is called a light year. Light travels very fast, indeed. In just one second it goes 186,000 miles.

Astronomers use light years to measure the distance from the earth to the stars. The nearest visible star, Alpha Centauri, is 4.3 light years or 26 trillion miles away. And the Great Spiral Nebula in the Constellation Andromeda, the farthest object in space that we can see without a telescope, is 1,500,000 light years from earth. This means that the light from the Spiral Nebula has been traveling for one and a half million years before reaching us.

After all those years, how can we be sure that the Spiral Nebula is still there? The answer is: we can't. In fact, the Spiral Nebula might have disappeared yesterday, and nobody would know it until one and a half million years from now!

WHAT DO B.C., A.D., AND B.P. MEAN?

Most people measure time by important happenings. Long ago, Hebrew religious leaders said that the most important event in their history was the creation of the world. And so the dates on their calendar showed the number of years they believed had passed since the creation.

Today, according to the Jewish religious calendar, the year when men first landed on the moon is written 5729.

Among Christians, the birth of Jesus was the most important date. They often referred to Jesus as "our Lord." And everything that happened after his birth was said to have happened "in the year of our Lord." This phrase was written *Anno Domini* in the Latin language, and it was then shortened to the initials, A.D. Although experts disagree about the exact year of Jesus' birth, Christian priests fixed a time which became the year A.D. 1.

According to this calendar, men first landed on the moon in 1969.

Of course, important events happened before A.D. 1. Christians who studied history figured out how to make dates from other calendars correspond to their calendar. All dates before A.D. 1 were dates Before Christ — that is, B.C. People who are not Christians often write it B.C.E., meaning *B*efore the *C*ommon *E*ra. It is called the Common Era because the Christian calendar is now commonly used in most parts of the world.

People who follow the Islamic religion (religion of Muslims) fixed their most important date in another way. They count years from the time when their leader Mohammed left the Arabian city of Mecca.

According to the Islamic calendar, men landed on the moon in the year 1347.

Some countries now use both the Christian calendar and their own calendar. Their holidays and histories are dated according to their own customs. But business letters to foreign countries are dated like European and American letters.

Most people don't spend much time on ancient history. Those who do often think that the A.D. and B.C. dating system is a nuisance. Suppose that, in the year 1980, an archeologist found out that traffic policemen, in 680 B.C., kept men from driving horses too fast on Chinese roads. How long ago would that be? To get the answer, it is necessary to add 1980 and 680 — 2,660 years ago. The archeologist who discovered an ancient Chinese traffic ticket would probably say it was written in 2,660 B.P. The letters B.P. stand for *B*efore the *P*resent. Of course, this B.P. date would have to be changed every year. In 1981, it would be 2,661 B.P.

Chapter XII

Numbers in Your Life

1. Why do we need telephone numbers?
2. Why do telephone dials have both numbers and letters?
3. Can you call a friend without using a phone number?
4. Why do we need area code numbers?
5. How can a family have 3.8 people?
6. Do you grow the same number of inches every year?
7. When people say you are above average, what do they mean?
8. Can a doctor really tell how many red blood cells you have?
9. Why do most library books have numbers?
10. Why don't all library books have numbers?
11. Why do we have ZIP codes?

WHY DO WE NEED TELEPHONE NUMBERS?

Suppose you lived a hundred years ago, just after the telephone was invented. If you wanted to phone your friend John Smith, you had to call the operator first. The dial telephone hadn't been invented yet. In small towns, people had phone numbers, but seldom used them. To call a friend, you told the operator "John Smith, please," and she rang his phone. In those days a small-town operator knew everybody who had a phone. She knew exactly how to connect your phone to John Smith's.

In a large city, operators couldn't possibly know all the people who had phones. When you wanted to talk to Sally Jones, you told her number to the operator, who could find it on the board and make the connection.

In the beginning this system worked. Operators could handle all the calls. But before long, in some large cities, thousands of people had phones. Operators couldn't keep up with all the connections that had to be made. Then an inventor figured out an automatic way to connect one phone with another. Modern dial phones grew out of this invention.

Young men were telephone operators at this huge switchboard in New York City in 1879.

WHY DO TELEPHONE DIALS HAVE BOTH NUMBERS AND LETTERS?

Can you memorize this number?
621-5463
Most people can remember at least one number like this — their own telephone number. But years ago the phone company wasn't sure everyone would be able to remember big numbers. So they made up numbers that contained words and digits. Then, when dial phones were invented, dials had to have both numbers and letters. If your number was
Maryland 1-5463
you dialed the first two letters of Maryland and then the remaining digits: MA 1-5463. This certainly did make phone numbers easy to remember. Of course, the M hole on the dial is the same as the 6 hole, and the A is the same as the 2, so you really dialed 621-5463.

In some places you still find words as part of phone numbers, but in most areas of the United States telephone numbers are now simply seven digits. Even though letters are easier to remember, the phone company relies more and more on digits.

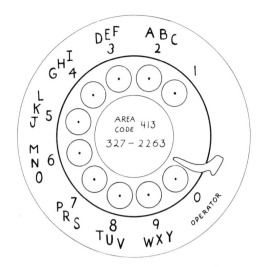

CAN YOU CALL A FRIEND WITHOUT USING A PHONE NUMBER?

Here's a trick that can sometimes help you remember a friend's telephone number without memorizing any numbers at all!

Suppose the number is 327-2263. First write it down the way it appears in the picture below.

Next, look at the telephone dial and copy under each numeral the letters which appear next to it on the dial.

Now look carefully at the letters. See if you can spell anything by taking a letter from each column. Sometimes you can figure out a word or words which, when dialed, are the same as your friend's number. For example:

```
3   2   7   2   2   6   3

D  (A)  P   A  (A)  M  (D)

(E)  B  (R) (B)  B  (N)  E

F   C   S   C   C   O   F
```

Now you don't ever have to remember 327-2263. If you want to call your friend, just dial EAR BAND.

Of course, this trick doesn't always work. A number such as 977-5554 is made up of all the wrong letters and can't be made into any words at all.

The phone company knows this trick, too. In New York City you can get a weather report by dialing WEATHER. And you can find out the time of day by dialing NERVOUS.

WHY DO WE NEED AREA CODE NUMBERS?

Phone numbers in some countries are seven numerals long. Seven numerals can be combined in different ways to give 10 million different phone numbers, beginning with 000-0000 and ending with 999-9999. This may seem like a great many numbers. But there are more than 140 million telephones in the United States — six million in New York City alone.

Ten million numbers aren't enough for 140 million phones. The phone company had a problem to solve, and this is how they did it: They divided the United States into more than 120 telephone areas, and gave each area a special three-figure number called an area code. This area code number was put in front of every regular number, increasing its length to 10 numerals. And now, with 10 numerals, 10 *billion* phone numbers are possible!

HOW CAN A FAMILY HAVE 3.8 PEOPLE?

An average family can have 3.8 people. What does this average family look like? Does the eight-tenths of a person sit down to dinner with everybody else?

Averages are made of numbers. Suppose you are going to find the size of the average family in a little village. You must count the number of families and the number of people living there. The village has five families. There are three people in the Fenwick family, two Gilligans, four Fongs, six Irvings, and four Cooks — a total of 19 people. Nineteen divided by five is $3^4/_5$ — or 3.8.

So we say the average family in this village has 3.8 people, even though such a family couldn't possibly exist.

DO YOU GROW THE SAME NUMBER OF INCHES EVERY YEAR?

You have been growing since you were born, but not always at the same rate. And you grew fastest before you were born. When you started growing, you were smaller than a pinhead. By the day of your birth, you were about 47 centimeters or 18½ inches long. You grew that much in less than a year. After you were born, your growth slowed down a bit. Around your fourth birthday your growth slowed down still more.

Children between the ages of five and ten usually grow at a steady rate. Some time after their tenth birthday, there is a big change. They grow very quickly for a short period of time. This happens to different children at different ages. Some girls have growth spurts when they are eleven, but most girls speed up when they are 12 or 13. Most boys have growth spurts between the ages of 13 and 15. When the spurt is over, growth slows down. People stop growing by the time they are in their twenties. After that, they actually shrink a little.

WHEN PEOPLE SAY YOU ARE ABOVE AVERAGE, WHAT DO THEY MEAN?

Suppose you are in a club with six other children, and you want to know if you are above or below the average height of the club members. You must calculate the average height and compare your height to this figure.

First, you find everyone's height, including your own. Next, add those numbers together. Then divide this sum by seven, the number of club members. The answer you get is the average height for your club. The chart below shows the heights of the club members. The average works out to be 57½ inches. If your height is 58 inches, you are slightly taller than average.

Would you like to be above average in everything? Probably not. You certainly would not like to have more than the average number of cavities in your teeth. No one is above or below average in everything. Everyone is above average in some things and below in others.

The sum of the heights is 402 inches, or 1,021 centimeters. Divide by 7 and you get 57 3/7 inches, or 146 centimeters, which is the average.

CAN A DOCTOR REALLY TELL HOW MANY RED BLOOD CELLS YOU HAVE?

The blood in your body is a liquid. The fluid part, called plasma, is straw-colored. The red color comes from tiny specks of living matter called cells. These red cells are so small they can't be seen without a microscope. They can't be measured in inches or centimeters. They are measured in a unit called a micron, which is one-thousandth of a millimeter. Each flat circular cell is about 8 microns wide — or about three ten-thousandths of an inch.

If you weigh 40 kilograms (about 88 pounds), you will have about 3 liters (3 quarts) of blood. Floating in your plasma, if you are healthy, are about 15,000,000,000,000 red cells. If they were all laid side by side, they would reach around the earth's equator almost three times!

The red cells are so important that you cannot live without them. They carry bits of oxygen to the other cells in your body. If you do not have enough red cells, something is wrong, and you feel sick. That is why a doctor is interested in finding out how many you have. Yet it is impossible to count them all, even if each one could be seen. Counting at the rate of 200 a minute, and counting all through a twenty-four hour day, it would take more than 150,000 years to count all the way up to 15,000,000,000,000! So what can be done?

First the doctor pricks your finger and collects a very small amount of blood. This is carefully measured. The blood is then mixed with a known amount of water. Next, a very tiny drop of the mixture is smeared over a piece of glass called a slide. This has been marked off in squares. Now the count can begin.

The doctor can look at the slide through a microscope and count the cells in each square. Since everything has been carefully measured, it is possible to figure out the number of cells in the whole sample. Suppose it shows there are 5,000,000 cells in one cubic millimeter of blood. Then this means you are healthy.

Not all doctors do blood counts. Most of them send samples to a laboratory where the counting may be done by an electronic instrument.

WHY DO MOST LIBRARY BOOKS HAVE NUMBERS?

Numbers make library books easy to find on the shelves. First you look at cards in the card catalogue and find one that gives the author's name and the name of the book you want. On the same card you may also see a number. The number is the clue that leads you to the shelf where that same number appears on the book.

This system seems obvious now. But a hundred years ago you had to be a real detective to find what you wanted in most libraries. Then along came Melvil Dewey.

Melvil loved efficiency and order. When he was fifteen years old, he decided to reform his mother's housekeeping. He persuaded her to let him make a new arrangement for everything she used in cooking. Nobody knows whether she really liked the new system, but Melvil thought it was good.

Later Melvil went to Amherst College in Mas-

Your Own Giant Numbers

You probably have about
120,000 hairs on your head
5,000,000 wrinkles, called villi, in the lining of the small intestine
3,000,000,000 tiny air sacs in your lungs
1,000,000 tiny narrow tubes in each kidney
100,000,000,000 cells in your brain
1,000,000,000,000,000 cells in your whole body, all supplied with oxygen which your 15,000,000,000,000 red blood cells carry through 60,000 miles of blood vessels.

sachusetts. To help pay his way, he worked in the college library. What he found there almost made him sick. The library was full of books, but they had simply been poked onto any convenient shelf. The librarian knew how to find things because he remembered where he had put them. That was not good enough for Melvil Dewey. Immediately he began trying to figure out how to number and arrange books so that anybody could find them easily.

First he queried people who worked in other libraries. Each librarian had his own system, but the systems all had faults. Dewey kept thinking — even in church while he pretended he was listening to the sermon. And one Sunday a plan popped into his head. Part of the plan came from good ideas other librarians had suggested. Part was Dewey's own invention. Put together, they made a simple, useful way of numbering library books.

Many librarians liked Dewey's plan, and they adopted it. They called it the Dewey Decimal System because Dewey really used a decimal point in numbering some books. For example, a book on the history of mathematics might be 510.09.

Other librarians have worked out entirely different systems for making books easy to find. But all of them use numbers, or numbers and letters, as clues.

WHY DON'T ALL LIBRARY BOOKS HAVE NUMBERS?

In most libraries books on certain subjects always have numbers, or numbers and letters. Science books, for instance, are numbered. So are books about such things as history, sports, painting. But books of short stories and novels are often marked only with the letter F, which stands for Fiction.

Now, suppose you want to borrow *The Hound of the Baskervilles* by A. Conan Doyle. You go to the fiction section. There the books are arranged alphabetically by the last names of authors. You look for the D-shelves, then for Doyle, then for *Hound* (because *The* is understood). Fiction books are easy to find in this way. And librarians can easily put them back on the shelves alphabetically. Numbers are not really needed.

But what about non-fiction books? Suppose you want a book on the history of chemistry. Would you look on the history shelf? Or on a science shelf? You decide on science. But there are many kinds of science. And there are several kinds of chemistry. Hunting through all the science shelves would take too much time. Even the librarian might forget exactly where the book belongs. So books that cause such complicated problems must have numbers.

WHY DO WE HAVE *ZIP* CODES?

One day somebody in Alabama mailed a letter that had only this address written on it:

Gid

80481

Two days later the mailman delivered the letter to the right person.

Letters and postcards like this one are called nixies in the United States. In a big post office a nixie goes to a special desk where a post-office sleuth figures out what the address might be. On this nixie the five-digit number was a clue. The sleuth decided it must be a ZIP code number. So he looked up the number in a special book and found it next to the name of a little town in Colorado. In such a small town, he said to himself, there was probably just one person called Gid, and the postmistress there would know who had that nickname. That was good detective work, and the ZIP code number made it possible for Gid to get the letter.

Every town in the United States has its own number, called a ZIP Code number. Big cities are divided into sections and each section has a ZIP. These code numbers weren't invented just for nixies. They help to make all the mail easy to deliver.

Before there were code numbers, some of the people who worked in post offices had to be memory experts. They had to remember where hundreds of different towns were located, so that sacks of mail would go into the right delivery trucks. They had to know the locations of hundreds of streets in big cities so that the postmen in each part of the city would get the right mail to deliver. Even a fast worker had to read the whole address on each letter and that took time.

A number can be read much more quickly, and all mail with the same number can be tossed immediately into the same sack. In some post offices there are machines that can recognize numbers, although they can't read whole addresses. Machines and people keep mail zipping along.

Towns and cities in many countries have their own codes. Some of them are more detailed than those in the United States. Some have both numbers and letters. When you address a letter to someone in the 14th section of the city of Helsinki in Finland, you have to write 0014 Helsinki 14. The number for Oakville in Ontario, Canada, is L6J 5E9.

Chapter XIII
Wonderful and Far Out

1. Would numbers exist if no one had ever thought of them?
2. What is the biggest number you can write with three digits?
3. What is infinity?
4. Why are some numbers called imaginary numbers?
5. Can a piece of paper have only one side?
6. How can you make something twice as big by cutting it in two?
7. Why is a Halloween mask like a pretzel?
8. "How many bulls' tails are needed to reach the moon?"
9. "How many black beans does it take to make three white beans?"
10. Can you learn to be a lightning calculator?
11. Which is shorter: an uphill mile or a downhill mile?
12. What is a time capsule?
13. Can we signal to people in other worlds?
14. What is the fourth dimension?
15. And now, our final question: Where did all these answers come from?

WOULD NUMBERS EXIST IF NO ONE HAD EVER THOUGHT OF THEM?

We learn to say the words for numbers and to write the symbols for them when we are quite young. We use these words and symbols so much that the numbers they stand for seem to be real things. We know that a shoe is real and that when we have two shoes, each of them is real. But what about the two? Where is it? What is it? It is not something that we can see or touch.

We cannot take a shovel and pile up twos as high as we want. Yet we can add twos together and get a number that is as large as we want. That takes thinking. Everything we do with numbers requires thinking, because numbers are ideas. And ideas exist only in people's minds.

WHAT IS THE BIGGEST NUMBER YOU CAN WRITE WITH THREE DIGITS?

If you do not know about exponents, you might answer 999. An exponent tells you to multiply a number by itself a certain number of times. Using exponents, you can write a much bigger number with three digits.

If the number five has an exponent of three, it means you must multiply five by five by five. Five with an exponent of three is written 5^3. To read it aloud, you say five raised to the third power. And 5^3 equals $5 \times 5 \times 5$ or 125.

It is possible for an exponent to have an exponent. Let's see what 5^{3^2} equals. First we calculate 3^2, which is 3×3 or 9. Then 5^{3^2} equals 5^9, or $5 \times 5 \times 5 \times 5 \times 5 \times 5 \times 5 \times 5 \times 5$, or 1,953,125.

$5 \times 5 = 25$
$5 \times 5 \times 5 = 125$
$5 \times 5 \times 5 \times 5 = 625$
$5 \times 5 \times 5 \times 5 \times 5 = 5,125$
$5 \times 5 \times 5 \times 5 \times 5 \times 5 = 15,625$
$5 \times 5 \times 5 \times 5 \times 5 \times 5 \times 5 = 78,125$
$5 \times 5 \times 5 \times 5 \times 5 \times 5 \times 5 \times 5 = 390,625$
$5 \times 5 \times 5 \times 5 \times 5 \times 5 \times 5 \times 5 \times 5 = 1,953,125$

The biggest three-digit number you can write is 9^{9^9}. That is nine multiplied by itself 387,420,489 times! Do you want to work that out? Before you start, you should know what you are getting into. Think about this: if you used numerals the size of the numerals in this book, you would need a piece of paper over 800 miles long to write the answer in lines the size of the lines in this book.

WHAT IS INFINITY?

After you learn to count, you can keep on counting as long as you like. If you think of an enormous number, you can always think of the next higher number. This means there can be no largest number. The counting numbers, one, two, three, and so on, go on endlessly. In other words, there is an infinity of counting numbers. There is also an infinity of fractions, and an infinity of decimal places. Can there be an infinity of snowflakes or germs? No. Infinity is only an idea that mathematicians invented. It cannot be applied to real things.

WHY ARE SOME NUMBERS CALLED IMAGINARY NUMBERS?

Suppose someone gave you a square vegetable garden and told you the garden had an area of exactly 36 square meters. The vegetables were growing well, but you noticed that rabbits were getting into the garden and eating them. When you decided to put a wire fence around the garden, it rained all day. You had to know how much fencing to buy, but you didn't want to walk around in the mud measuring the sides of the garden. How could you solve your problem without getting your feet muddy?

To get the length of a side, you must find a number that, when multiplied by itself, will equal 36. Six times six equals 36. So each side of the garden is six meters long.

We can also say that the square root of 36 equals six. This is written in symbols, $\sqrt{36} = 6$. The length of the side of any square is the square root of its area.

The square root of four is two because two times two equals four. And the square root of one is one because one times one equals one.

People have known about square roots for a long time. But then mathematicians discovered something strange. They found that the answer to some very complicated problems was the square root of minus one. This answer seemed impossible. It could not be used to measure anything, and it did not seem to fit with other numbers. One mathematician was so upset by the square root of minus one that he called it "meaningless, fictitious, impossible, and imaginary." He called it all sorts of bad names, but still he wrote it down and worked with it. The name *imaginary* stuck. It really is a silly name. After all, numbers are ideas, and all numbers come from people's imagination.

CAN A PIECE OF PAPER HAVE ONLY ONE SIDE?

The surprising answer is "Yes!"

Take a long, skinny strip of paper. Now bring the ends toward each other as if you were going to join them to make a round band. But before finishing the band, twist one of the ends halfway around. Then tape the two ends together. You now have something with a special name — a Moebius band. And a Moebius band is a piece of paper with only one side.

How do you know for sure it has only one side? Try this: Start drawing a line along the center of the Moebius band. Keep on drawing. Your line will go all the way around the paper, returning to where you started. And you will never have to lift your pencil. Now, no matter where you look on the band, it will have a line on it. This proves that the Moebius band has only one side — the side with the line on it!

HOW CAN YOU MAKE SOMETHING TWICE AS BIG BY CUTTING IT IN TWO?

First, make a Moebius band. Draw a line along the center all the way around. Now take a pair of scissors, slit the paper a bit along the line to get the points of the scissors in, and cut the band down the middle, along the line you have just drawn. What do you think will happen? Will you end up with two separate Moebius bands? The answer is "No!" You will get only one band, twice as big around as the original.

Does it have one side or two sides? Draw a line along its center and find out.

1. To make a Moebius band, first give a twist to one end of a strip of paper. Then fasten the ends together.

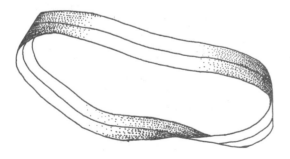

2. Draw a line along the center of the strip.

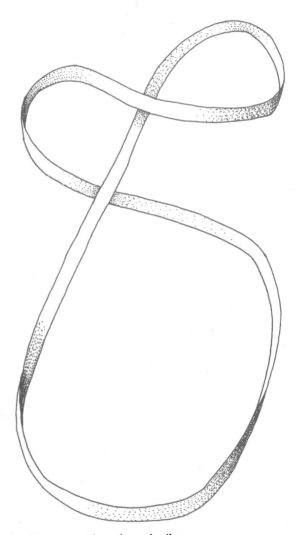

3. Now try cutting along the line.

WHY IS A HALLOWEEN MASK LIKE A PRETZEL?

The Halloween mask in the picture has three holes in it, and so does the pretzel. The mask and the pretzel don't look alike at all, but there are some people who say they really are the same! These people aren't joking. They belong to a special group of mathematicians who are called topologists.

A topologist says that two things are alike if you can pull, stretch, or squeeze one of them so that it looks exactly like the other one. No tearing is allowed, and you can't poke any new holes or pinch shut any old ones. Everybody, even a topologist, knows that a pretzel will crumble if you try to stretch it into a Halloween mask. So topologists pretend that everything is made of soft putty.

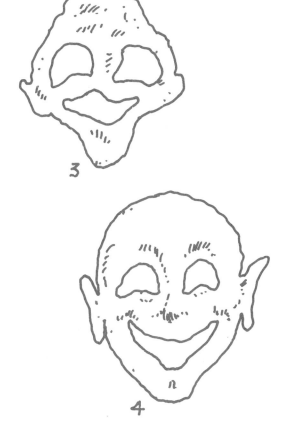

Steps in changing a putty pretzel into a mask.

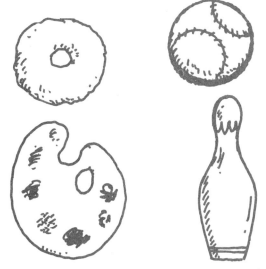

A doughnut and a pallet are alike to a topologist. So are a baseball and a bowling pin. Do you see why?

Now take your "putty" pretzel, and shape it into a mask. Stretch the top to make a forehead. Pull the middle out to make a nose and cheeks. And pull the bottom down to make a chin.

Topology, the branch of mathematics that topologists study, sounds very strange. But topology can be very useful. It helps scientists design the tiny electronic circuits that make up the insides of computers. Topology can help mapmakers make maps, and it can help knitters devise new stitches.

"HOW MANY BULLS' TAILS ARE NEEDED TO REACH THE MOON?"

Somebody yelled that question at twelve-year-old George Bidder. George stood at the front of a room full of people. All of them had bought tickets, just to come and ask him questions. Most of the audience believed George could solve very hard arithmetic problems very quickly in his head. They called him a lightning calculator. Others thought he was faking, and they tried to tease him. So one man asked, "How many bulls' tails are needed to reach the moon?"

"One, if it's long enough!" George called back.

Everybody laughed. Then serious questions went on.

"Suppose you have a wagon wheel that is 5 feet 10 inches around. How many times will it turn if it travels 800,000,000 miles?"

George thought for less than a minute. Then he gave the right answer: "724,114,285,704 times, with 2/7 of a turn left over."

George was not a fraud. He could solve even harder problems without using pencil and paper. At that time, in the year 1818, pocket calculators hadn't been invented. No one had ever taught George how to do arithmetic. When he was six years old he learned in school to count up to 100. He didn't learn to read and write. But numbers fascinated him. Instead of writing down figures, he arranged pebbles in rows. Then, by counting the pebbles, he got the answers to problems that he made up. Before long he could get the answers without the pebbles, just by thinking about his problems.

The neighbors loved to quiz George. Sometimes he would go to the blacksmith shop in the little English village where he lived, and the men who gathered there would give him hard problems to solve. For each right answer he got a penny.

Soon people in other towns heard about George. He and his father began traveling, putting on shows. The boy's skill seemed very strange and mysterious. When college professors examined him, they could not explain how he did such complicated arithmetic. George himself could not tell them. The answers just suddenly came to him.

When George Bidder grew up, he was still a very good calculator. He learned to read and write and became an engineer. But he still could not tell exactly how he got his answers. He knew that he never thought about numbers as if they were written down. He said that he imagined them in groups, almost as if they were pebbles arranged in rows, the way he did when he was a little boy. To this day nobody knows how he could solve problems so much more quickly than other people.

"HOW MANY BLACK BEANS DOES IT TAKE TO MAKE THREE WHITE BEANS?"

"Three, if you skin them," young Zerah Colburn answered.

Zerah was a lightning calculator, like George Bidder. He, too, put on shows, and sometimes people asked him silly questions for fun. But that didn't bother him. When he was eight years old he could give the answers to long, hard problems faster than anyone could write down the figures in the answers.

Zerah could not fully explain his strange power. He did know how he solved some problems. But even when he explained his method, it was hard to understand. For example, when he multiplied 4,395 by 4,395, he said he multiplied 293 by 293 by 15 by 15. He could not tell how he knew that this combination would give him the right answer.

Zerah Colburn was born in Vermont in 1804. Another famous calculator named Thomas Fuller was born in Africa. When Fuller was a boy, slave traders kidnapped him and sold him to a farmer in Virginia. He never went to school or learned to read or write, but he was a genius at arithmetic. Once he was asked how many seconds there are in 70 years, 17 days and 12 hours. At first his answer seemed to be wrong. But it wasn't. The man who asked the question, and had figured it out on paper, had forgotten that there are extra seconds in leap years.

A German boy named Karl Friedrich Gauss grew up to be one of the world's greatest mathematicians. When he was six years old, his teacher told him to add up all the numbers from 1 through 100. He did this so quickly that his teacher thought he might be cheating. But Gauss had simply figured out a shortcut in calculation. Instead of adding up a long column of 100 numbers, he added 1 to 100 and multiplied the sum by 50.

You can see why this works if you start out with an easier problem: "Add up all the numbers from 1 through 10."

Write the numbers the way you see them below in two columns. Add them across and then add up the sums this way:

$$\begin{align} 1 + 10 &= 11 \\ 2 + 9 &= 11 \\ 3 + 8 &= 11 \\ 4 + 7 &= 11 \\ 5 + 6 &= \underline{11} \\ & 55 \end{align}$$

As you can see, there are 5 pairs of numbers, and each pair adds up to 11. So, instead of adding the five elevens, you multiply: $11 \times 5 = 55$.

If you write the numbers 1 through 100 the same way, you will have 50 pairs, each pair adding up to 101. Young Gauss saw this and got the right answer: $101 \times 50 = 5050$.

CAN YOU LEARN TO BE A LIGHTNING CALCULATOR?

Very few people can do enormous arithmetic problems in their heads. It is not a skill that can be learned. But you *can* learn tricks that will astonish anyone who doesn't know your secrets. Here is one:

You say, "Give me a two-digit number that ends in 5."

"45," someone says.

"Now I can multiply 45 by 45 just as fast as I can write down the answer," you boast.

This is the trick: In your head you multiply 4, which is the first digit in forty-five, by 5, the next higher number. That makes 20. Write down 20, and after it write 25. That's the answer: 2,025. Now check the answer on paper or on a hand calculator. It is 2,025.

Try it again with 65 times 65. Multiply 6 by 7 in your head. That makes 42. Put down 42, then 25, making 4,225. Correct.

The trick works with any two-digit number that ends in 5. It seems like magic, but you can prove that it always works.

Karl Friedrich Gauss.

A 1975 automobile is one of the things a man in Nebraska is preserving in a "time capsule" of his own making under his front yard.

WHICH IS SHORTER — AN UPHILL MILE OR A DOWNHILL MILE?

Starting at the foot of Pike's Peak, a car travels 18 miles to the top. Does it travel the same number of miles coming down? Of course it does, if you use any kind of modern measuring system. In the metric system, the distance is 28.98 kilometers going up and 28.98 kilometers coming down.

But is this kind of measurement the only way to think about distance? The ancient Chinese believed there was another way. Suppose, they said, you are walking uphill carrying a heavy load. The distance *feels* much longer on the way up than it does on the way down.

Perhaps it feels as if you are going two miles uphill but only one mile down. Now, is it fair to say the uphill mile is exactly the same as the downhill mile? The Chinese would have answered "No." You must call the uphill distance two miles because it *feels* like two miles. So you must write the problem out like this:

2 miles uphill = 1 mile downhill = 5,280 feet
1 mile uphill = ½ mile downhill = 2,640 feet

So you can see that an uphill mile is shorter than a downhill mile.

WHAT IS A TIME CAPSULE?

Can you communicate with the distant future? It sounds impossible, but people have found a way to do it. They put records and messages into the cornerstones of public buildings. If a building is torn down, years later, workmen will find the messages from the past.

Records, messages, souvenirs, and pictures are also put into special containers called time capsules. Usually a time capsule is made of metal and filled with nitrogen gas which keeps the contents from decaying. After it is carefully sealed, it is buried underground. A plaque marks the burial place and gives the date when the capsule should be opened. This can be 100, 1,000, or even 10,000 years in the future.

The first time capsule was buried 50 feet underground at the site of the 1939 New York World's Fair. It contains records, photographs, and other souvenirs. The capsule is to be opened in the year 6939, long after most other records of the 20th century have disappeared.

Sometimes people put peculiar things in time capsules. In 1976, to celebrate the United States Bicentennial, the state of Massachusetts put a time capsule in a room in the state capitol to be opened in 2076. When it is opened, people will find the usual collection of official documents. But they will also find a tennis ball, a razor, a package of instant chicken soup, a jar of coffee, and two cans of cranberry juice. To top it off, they will find a pair of size 13 shoes that were once worn by a Massachusetts legislator!

CAN WE SIGNAL TO PEOPLE IN OTHER WORLDS?

Is anybody listening out there in space? Are there people in other worlds who can receive radio messages?

Many scientists believe living creatures may exist elsewhere in the universe. Perhaps they are just as smart as we are. Maybe they have even discovered radio broadcasting and receiving. Of course, they would not be likely to speak any of earth's languages. So how would they understand a message if we could broadcast it to them? Is there any language we could use?

Mathematicians say, "Yes! Anyone who can invent radio must understand numbers." Perhaps they, would get the idea if we simply sent out signals as if we were counting: beep, beep-beep, beep-beep-beep. But that might be too simple. Perhaps we should send complicated signals that would prove we really do know complicated mathematics. And that is just what scientists have done. They have broadcast signals that show they can multiply 23 by 79.

Why those two numbers? They were chosen because they belong to a group called *prime numbers*. A prime number is one that can be divided only by itself and by one. Anyone with a radio in another world would surely know about prime numbers and would recognize the signal.

The station that broadcasts the signals is on the island of Puerto Rico. It can also receive signals from far away if any station in space tries to reach us. The only trouble is that it will take years and years to get an answer to 23 times 79.

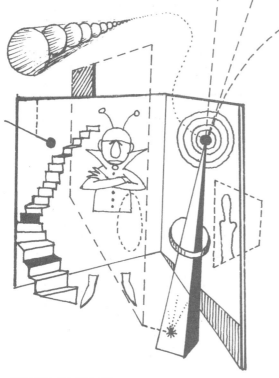

WHAT IS THE FOURTH DIMENSION?

Two-dimensional things are flat. Table tops, sheets of paper, and floors all have two dimensions — length and width. Three-dimensional things have height as well. Tables, pads of paper, and rooms are three-dimensional. Mathematicians sometimes think of a corner of a room with two walls and a floor as a working model of three-dimensional *space*. The space is made up of all the points in the room. You can locate any point in the room if you know exactly how far it is from each of the two walls and the floor.

Three-dimensional space makes sense. But what is four-dimensional space, and what does it look like?

When mathematicians first started thinking about these questions at the beginning of the last century, many people were suspicious.

"How can you talk about things you can't see?" they asked.

The mathematicians had an answer.

"Mathematics is a collection of ideas, and ideas don't have to look like anything in particular."

Years passed. Finally people began to understand what four-dimensional space must be like. It is like a corner of a room with three walls instead of two. But the new wall is not just another ordinary wall of the room. It isn't the ceiling, either, and in fact it is not like a wall anyone has ever seen. The new wall goes in an entirely different direction from the other walls, and this direction is almost impossible to imagine. But not quite impossible. Mathematicians can imagine it, even though they all agree that the new wall is impossible to see.

Why did they add the new wall? Because they wanted to locate the points of four-dimensional space. In three-dimensional space you can locate a point by finding its exact distance from the walls and the floor. The extra wall in four-dimensional space means you can locate a point in this space, too, once you know its distance from the three walls and the floor. Mathematicians begin to understand a space when they know how to locate its points.

Then people asked, "If this is what four-dimensional space is like, can there be a five-dimensional-space, and what does *it* look like?" Again, mathematicians had an answer.

"Add another wall which goes in still another entirely different unimaginable direction."

After that, the walls kept coming. Soon people began thinking about an infinite space with an infinite number of walls, all going in totally different unimaginable directions. And once people had thought of an infinite space, mathematics became truly "wonderful and far out."

Albert Einstein developed new mathematical ideas using the fourth dimension.

AND NOW, OUR FINAL QUESTION:

WHERE DID ALL THESE ANSWERS COME FROM?

No single one of us who worked on this book would ever have been able to answer all these questions. There were lots of things we didn't know. In some cases we found our facts in magazines and newspapers. We are especially indebted to Professor Gerald Oster's article on starfish in *Natural History*. Other information came from people who work in different industries and government agencies. We are particularly grateful to the United States National Bureau of Standards and to the American Numismatic Society for their help.

For facts about the past we have turned to books on the history of science and mathematics.

These were our main references:

A History of Mathematical Notations, by Florian Cajori

Number, the Language of Science, by Tobias Danzig

Let's Go Metric, by Frank Donovan.

In Mathematical Circles, by H. W. Eves

ESP, a Scientific Evaluation, by C. E. M. Hansel

Mathematics in Western Culture, by Morris Kline.

Number Words and Number Symbols, by Karl Menninger

The Exact Sciences in Antiquity, by Otto Neugebauer

Lost Discoveries, by Colin Ronan

History of Mathematics, by D. E. Smith

Science Awakening, by B. L. Van der Waerden

The four of us have enjoyed gathering all this information, and we hope that as you read it, you will enjoy it, too.

Mary Elting
Rose Wyler
Eva-Lee Baird
Robert Moll

Index

A

Abacus, 58, 59
Adder, 69
Adding machine, 59
Africa, 16, 86, 101, 121
African gaming pieces, 86
Algebra, 62, 63
    WHAT IS THE DIFFERENCE BETWEEN ARITHMETIC AND ALGEBRA?, 62
Alpha Centauri, 106
Amateur athlete, 30
American Revolution, 17, 79
Amherst College, 113
Amicable numbers, 11
Analytical engine, 64
Angle(s), 88
    WHY ARE THERE RIGHT ANGLES BUT NO LEFT ANGLES?, 88
Anker, 77
Anno Domini, 106
Arab(s), 45, 46, 47, 60
Arabic numerals, 45, 46, 47, 58
Arch, 82
Area code, 110
Arithmetic, 62
Asia, 16, 45
Astrology, 34-35
Astronaut(s), 50, 82, 83
Astronom(er)(y), 35, 40
Australian crawl, 32
Automobile(s), 96-98
    HOW CAN YOU FIGURE MILES PER GALLON?, 96
    HOW DO YOU READ A MILEAGE TABLE?, 97
    HOW FAR CAN A CAR TRAVEL ON A GALLON OF GAS?, 96
    IS A SPEEDOMETER THE SAME AS AN ODOMETER?, 97
    WHAT DOES A "MEASURED MILE" MEAN?, 98
    WHY ARE THERE BOTH LETTERS AND NUMBERS ON SOME LICENSE PLATES?, 97
    WHY CAN YOU GO FARTHER ON A GALLON OF GAS IN CANADA THAN IN THE UNITED STATES?, 96
Average(s), 111
    batting, 28
    family, 110
    law of, 12
    HOW CAN A FAMILY HAVE 3.8 PEOPLE?, 110
    WHEN PEOPLE SAY YOU ARE ABOVE AVERAGE, WHAT DO THEY MEAN?, 111
Avoirdupois pound, 76

B

Babbage, Charles, 64
Babylonia, 34, 47, 56, 100
Baker's dozen, 91
Banana(s), 88
Banca, 20
Bank(s), 19, 20, 25, 48, 68
    HOW DID BANKS GET STARTED?, 19
Bankrupt, 20, 24
    WHAT DOES IT MEAN TO GO BANKRUPT?, 24
Bannister, Roger, 28
Baseball, 28-29
Batting average, 28-29
Bayi, Filbert, 28
Belgium, 9
Bicentennial, 122
Bidder, George, 120
Big Dipper, 103
Binary system, 55
Bird(s), 50, 82
Blackburn, Douglas, 39
Blink, 77
Blood, 112
Bodge, 77
Body, 112
Bomb calorimeter, 80
Bow, 77
British Imperial System, 78
Buddhist, 101
Budget, 93
Bug, computer, 65
Butt, 77

C

Cab, 77
Calculation, 46
Calculator, pocket, 59, 61, 69, 72
    IS THERE ANY DIFFERENCE BETWEEN A CALCULATOR AND A COMPUTER?, 69
    HOW DOES A POCKET CALCULATOR WORK?, 69
    WHY DO THE NUMBERS ON A POCKET CALCULATOR LOOK SO STRANGE?, 72
Calendar, 100-102, 105, 106
    WHY DO JEWISH HOLIDAYS COME ON DIFFERENT DATES FROM YEAR TO YEAR?, 102
    WHY DOES THE CALENDAR HAVE TWELVE MONTHS? 100
    WHY ISN'T EASTER ON THE SAME DATE EVERY YEAR?, 102
Calories, 80
Canada, 17, 96, 114
Cape Verde Islands, 105
Card, 94
Catty, 77
Cavil, Charles and Syd, 32
Cells, red blood, 112
Celsius, Anders, 80
Celsius scale, 48, 80
    WHY DO THERMOMETERS HAVE DIFFERENT SCALES?, 80
Census, 41
Centimeter, 78
Chain letter(s), 14-15
Charlemagne, 51
Checkers, 69, 70
Chess, 67
China, 9, 37, 106, 122
Christian(s), 60, 101, 102, 106
Circle, 85, 86
Clemens, Samuel, 75
Clock arithmetic, 62
Clove, 77
Coin(s), 16, 17, 19, 77, 104
Colburn, Zerah, 121
Comet, 36
Comet pills, 36
Computation, 58-72
    CAN YOU MULTIPLY JUST BY ADDING?, 61
    DOES 1 + 1 ALWAYS EQUAL 2?, 55
    HOW CAN 10 + 5 = 3?, 62
    HOW CAN YOU ADD WITHOUT WRITING DOWN NUMBERS?, 58
    WHERE DID THE SIGN FOR "DIVIDE" COME FROM?, 60
    WHERE DID THE SIGN FOR "EQUAL" COME FROM?, 60
    WHERE DID THE SIGN FOR "MULTIPLY" COME FROM?, 60
    WHERE DID THE SIGNS FOR "PLUS" AND "MINUS" COME FROM?, 59
    WHICH IS FASTER: AN ABACUS OR AN ADDING MACHINE?, 59
    WHO WERE THE ZERO FREAKS?, 47
Calculator, lightning, 120-121
    CAN YOU LEARN TO BE A LIGHTNING CALCULATOR?, 121
Computer(s), 41, 55, 60, 64-71
    address, 67
    bug, 65
    checkers playing, 69
    chess playing, 67
    memory, 66
    pi, 83
    program, 65, 69, 70
    CAN A COMPUTER MAKE YOU RICHER?, 68
    CAN A MACHINE PLAY CHESS?, 67
    CAN COMPUTERS THINK?, 70
    HOW DOES A COMPUTER'S MEMORY WORK?, 66-67
    WHAT IS A BUG IN A COMPUTER PROGRAM?, 65
    WHAT IS A COMPUTER PROGRAM?, 65
    WHO INVENTED THE COMPUTER?, 64
Constellation, 34-35, 106
Count-down, 50
Counting, 44, 46-51, 56
    on fingers, 71
    CAN A DOCTOR REALLY TELL HOW MANY RED BLOOD CELLS YOU HAVE?, 112
    CAN ANY ANIMALS COUNT?, 50
    CAN YOU COUNT WITHOUT NUMBERS?, 48
    DO ALL PEOPLE COUNT THE SAME WAY?, 56
    DO YOU REMEMBER HOW YOU LEARNED TO COUNT?, 49
    DID PEOPLE EVER COUNT ON THEIR TOES?, 49
    WHY DO ASTRONAUTS USE "COUNT-DOWN" INSTEAD OF "COUNT-UP"?, 50
    WHY DO WE COUNT EGGS BY THE DOZEN?, 51
    counting board, 46
Cran, 77
Credit card, 94
Crith, 77
Cube, 82, 86
Cubit, 74

D

Day, 103
Dead Letter Office, 15
Debug, 65
Decimal(s), 52, 55, 83
    WHAT ARE DECIMALS?, 52
    WHO INVENTED DECIMALS?, 52
Degree(s), 80, 88
Depth, measure of, 75
Descartes, Rene, 63
Dewey, Melvil, 112-113
Dewey Decimal System, 113
    WHY DO MOST LIBRARY BOOKS HAVE NUMBERS?, 112
    WHY DON'T ALL LIBRARY BOOKS HAVE NUMBERS?, 113
Dial telephone, 108-110
Diameter, 83
Dice, 86
Difference Engine, 64
Digit(s), 71, 94, 116
Digital clock, 71
Digital computer, 71
    WHAT HAS DIGITS BUT NO HANDS?, 71
Divide, sign for, 60
Dodecahedron, 84
Dog, 50, 86
Dollars(s), 9, 17, 18, 26
    bill, 9
    sign, 17
Dome, 82
Double-digit inflation, 94
Downhill mile, 122
Doyle, A. Conan, 113
Dozen, 51, 91
Dunninger, Joseph, 38

E

Eagle bone, 100
Earband, 109
Eclipse, 40
Edward I. King, 74
Egg, 31, 51, 82
Egypt, 9, 16, 22, 34-35, 59, 85, 86, 100, 103
Einstein, Albert, 124
Election(s), 42
Elizabeth I, Queen, 98
Ell, 77
England, 10, 17, 18, 19, 28, 31, 71, 76, 78, 91, 93, 98

Equal sign, 60
Equator, 53, 79, 112
Equinox, 102
ESP, 38-39
Exponent(s), 116
Extrasensory perception, 38
   What Is ESP?, 38-39

F
Fahrenheit, Gabriel Daniel, 80
Fathom, 75
Fathometer, 75
Feathers, 76
Fifty-fifty chance, 10
Finger-counting, 71
   What Was the First Digital Computer, 71
Fingerbreadth, 78
Firkin, 77
Flood(s), 34, 85
Florence, 47
Foot, 74, 78
Football, 30
Fortunetelling, 34-35, 40
   Can a Mind Reader Really Read Your Mind?, 38
   How Did Astrology Begin?, 34-35
   How Magical Are Magic Squares?, 37
   Is a Comet an Evil Omen? 36
Fourth dimension, 124
   What Is the Fourth Dimension?, 124
Fraction(s), 44, 51, 83
   Who Invented Fractions?, 51
France, 63, 78, 92, 93
Fuller, Thomas, 121
Furlong, 98

G
Gallon, 78, 96
Gaming pieces, 86
Gasoline, 96
Gauss, Karl F., 121
Geometry, 85
Germany, 52, 60, 74, 92, 104
Gold, 16, 76
Gold Rush, 16
Goldsmith, 19
Gram, 78
Gravity, 36
Great Spiral Nebula in Andromeda, 106
Greek(s), 11, 19, 34, 35, 36, 45, 51, 54, 74
Growth, 88, 111
   Do You Grow the Same Number of Inches Every Year?, 111
Gudea, 74

H
Haegg, Gunder, 28
Hairbreadth, 78
Hairbreadth escape, 78
Halley, Edmund, 36
Halley's comet, 36
Halloween mask, 119
Hands, The, 59
"Heads and tails," 12, 19
Hebrew(s), 36, 45, 101, 102, 106
Henry III, King, 18
Hexagon, 85
Hide, 77

Hindu numerals, 45
Holy fourfoldedness, 54
Horizon, 82
Horoscope, 35
Hour(s), 103, 104, 105
Huckleberry Finn, 75

I
Imaginary numbers, 117
Imperial gallon, 96
Inch, 74, 78
Income tax, 22
Indian(s), 8, 44, 56, 100
Indiana, 83
Infinity, 117
   What Is Infinity?, 117
Inflation, 94
Interest, 25, 48, 68
International Date Line, 105
Islam(ic), 45, 106

J
Japan, 8, 59, 104
Jehovah, 37
Jesus, 60, 106
Jet lag, 104, 105
   What Is a Great-Circle Route?, 104
   What Is Jet Lag?, 104
Jew(s)(ish), 37, 101, 102, 106
Joachimsthaler, 17

K
Kell, George, 29
Kempelen, Baron Wolfgang von, 67
Kilogram, 78
Kilometer, 78, 122
Kilowatt-hour(s), 92
Knickerbocker Base Ball Club, 29
Knuckle bones, 86
Krypton-86, 79

L
Lagash, 74
Lambert, Johann Heinrich, 83
Length:
   of day, 103
   of foot, 74
   of meter, 79
   of mile, 98
   of year, 100
Library, 112-113
License plate, 97
Light, speed of, 53
Light year, 106
Lightning calculator(s), 120-121
   Can You Learn To Be a Lightning Calculator?, 121
   "How Many Black Beans Does It Take To Make Three White Beans?", 121
   "How Many Bulls' Tails Are Needed To Reach the Moon?", 120
Lincoln, Abraham, 66
Liter, 78
Loaded dice, 86
Loan shark, 23
Log, 61, 77
Logarithm(s), 61
Long shot, 10
Lottery, 12
Louse, 78
Louse egg, 78

Love, 31
Luck(y), 8-12, 34, 36
   Are There Any Lucky Numbers?, 8
   Are There Any Lucky Shapes?, 84
   Do People Ever Have Lucky Streaks?, 8
   Does It Pay To Keep On Buying Tickets In A State Lottery?, 12
   Is 13 An Unlucky Number?, 9
   Were You Born Under A Lucky Star?, 34
   What Is a Fifty-Fifty Chance?, 10
   What Is a Long Shot?, 10
   What Is the Law of Averages?, 12
   Would You Believe It?, 11

M
Magic, 38
Magic squares, 37
Magellan, Ferdinand, 105
Mana, 104
Mars, 37
Mass, 76
Maya numerals, 56
Measure(ment)(s), 74-80, 85, 91
   distance, 97, 98, 106, 122
   time, 100-106
   Can a Fat Man Reduce By Going to the Top of a Mountain?, 76
   How Can the Ocean's Depth be Measured?, 75
   How Can You Lose a Day?, 105
   How Do We Know How Many Calories Are in a Hamburger?, 80
   How Do We Know the Length of a Meter?, 79
   How Do We Know the Length of the Year?, 100
   How Do You Read an Electric Meter?, 92
   How Far Away is the Horizon?, 82
   How Far Is It to the Bottom of the Sea?, 75
   What Do B.C., A.D., and B.P. Mean?, 106
   What Is a Light Year?, 106
   What Is a Metric Mile?, 28
   When Was a Foot Not a Foot?, 74
   When Was a Mile Not a Mile?, 98
   Which Weighs More—A Pound of Feathers or A Pound of Gold?, 76
   Why Are There 60 Minutes In an Hour?, 104
   Why Are There 24 Hours In a Day?, 103
   Why Is 13 a Baker's Dozen?, 91
Memory, computer, 66, 67
Meter, 28, 78, 79

Meter, electric, 92
Metric mile, 28
Metric system, 78, 79
   What Is a Hairbreadth Escape?, 78
   Why Didn't the U.S. Adopt the Metric System When Other Countries Did?, 79
Mexico, 8, 17
Mile(s), 78, 96, 97, 98, 122
   four-minute, 28
   metric, 28
   Which is Shorter—An Uphill Mile or a Downhill Mile?, 122
Miles per gallon, 96
Mileage table, 97
Milk carton, 91
Millimeter, 78
Million dollars, 26
Mind reader, 38-39
Mindinao Deep, 75
Minimum wage, 22
Mint, 19
Minus, 59
Minute(s), 101
Moebius band, 118
Mohammed, 45, 106
Money, 14-26, 93
   Can Making Money Cost Too Much?, 26
   Can You Make a Fortune With Chain Letters?, 14
   How Could You Get Rich Quick?, 14
   How Does a Credit Card Work?, 94
   Is a Loan Shark an Animal?, 23
   Is Anything Cheaper By the Dozen?, 91
   What Do Numbers on a Milk Carton Mean?, 91
   What Does a Million Dollars Look Like?, 26
   What Is a Budget?, 93
   What Is a Minimum Wage?, 22
   What Is a Profit?, 25
   What Is a Sales Tax?, 92
   What Is an Income Tax?, 22
   What Is Double-Digit Inflation?, 94
   What Is Pin Money?, 93
   When Is a Penny Not a Penny?, 18
   Where Did the Dollar Sign Come From?, 17
   Where Did the Pound Sign Come From?, 19
   Where Does the Word Dollar Come From?, 17
   Why Don't All People Use the Same Kind of Money?, 16
Money-changer(s), 19-20
Monkey gaming pieces, 86
Moon, 82, 100, 101, 102
Moon-calendar, 101, 102
Multiply, 60, 61
Muslim(s), 37, 45, 106

N

Nail, 77
Napier, John, 52, 61
Negative number, 48
New Zealand, 28
Newton, Isaac, 36
Nile River, 34, 85
Nit, 77
North Pole, 79, 104
Nox, 77
Noy, 77
Number(s), 8, 11, 19, 36, 44-56, 58, 116-117, 123
    Are There Any Triangular Numbers?, 54
    calculator, 69, 72
    of red cells, 112
    imaginary, 117
    How Did Names For Numbers Get Started?, 44
    library book, 112, 113
    license plate, 97
    prime, 123
    telephone, 66, 108-110
    What Is a Round Number?, 53
    What Is a Square Number?, 53
    What Is the Biggest Number You Can Write With Three Digits?, 116
    Which Came First—Zero or One?, 47
    Why Are Some Numbers Called Imaginary?, 117
    Why Do We Call Numbers a Great Invention?, 44
    Why Is Pi Such a Strange Number?, 83
    Would Numbers Exist If No One Had Ever Thought of Them?, 116
number line, 48
Numeral(s), 45
    Arabic, 45, 46, 47
    Hindu, 44, 45
    Maya, 56
    Roman, 45, 46, 47
    How Did the Romans Add With Roman Numerals?, 46
    Where Did Our Numerals Come From?, 44-45

O

Octagon, 85
Odometer, 97, 98
Omen, 36
One-dollar bill, 9, 26
Other worlds, 123
Oughtred, William, 60
Ounce(s), 76, 78

P

Parallelogram, 85
Parapsychology, 38-39
Pellos, Francesco, 52
Pence, 18
Penny, 12, 18
Pentagon, 84
Pentagram, 84
Perch, 77
Perfect number, 8
Persian, 44
Peso, 17
Philippine Islands, 75
Phone, *see* telephone
Pi, 83
Pigafetta, Antonio, 105
Pin money, 93
Pint, 78
Plus, 59
Pocket calculator, *see* calculator
Polygon, 85
Poll(s), 42
Positive number, 48
Pound(s), 78
    avoirdupois, 76
    English, 19
    troy, 76
Pound sign, 19
Precognition, 39
Prediction(s), 34, 40, 41
    Are There Any Real Ways To Predict the Future?, 40
    Can Cards Tell Your Future?, 40
    How Do Public Opinion Polls Work?, 42
    What Is the Census?, 41
Pretzel, 119
Price, 90, 91
Prime numbers, 123
Professional athlete, 30
Profit, 25
Psychics, 39
Public opinion polls, 42
Puerto Rico, 123
Punched card, 94
Pythagoras, 84
Pythagorean(s), 54, 84

Q

Quart, 78
Quarter Horse, 30
Quartz, 71
Queensland, 56

R

Race(s), 28-29
Radio broadcasting, 123
    Can We Signal To People In Other Worlds?, 123
Recorde, Robert, 60
Religion(s), 60, 101, 102, 106
Rhine, Joseph, 39
Riddle, 63, 72
Riese, Adam, 60
Right angle, 88
Right triangle, 85
Rocket, 50
Rome(an), 8, 9, 19, 46, 47, 93, 98
Roman numerals, 45, 46, 47
Rotl, 77
Round number, 53
Rudolff, Christoff, 52

S

St. Andrew's cross, 60
Sales tax, 92
Savings bank, 25
Second(s), 104
Seer, 77
Seven Gods of Good Luck, 8
Seventh son, 8
Shapes, 82-88
    Can a Piece of Paper Have Only One Side?, 118
    Could a Bird Sit on a Square Egg?, 82
    How Can You Make Something Twice As Big By Cutting It In Two?, 118
    How Many of These Shapes Can You Name?, 85
    Why Does the Earth Look Flat When It Is Really Round?, 83
    Why Is a Halloween Mask Like a Pretzel?, 119
Shark, 23
Shekel, 77, 104
Shid, 77
Siam, 78
Silver coins, 16
Simple interest, 68
Sirius, 34
Smith, G. A., 39
Snail, 88
Solar calendar, 102
Solar year, 100
South Pole, 101
Space, 123, 124
Spain, 92
Spat, 77
Speedometer, 97
Spiral nebula, 106
Sports:
    What Is a Batting Average?, 28-29
    What Is a Quarter Horse?, 30
    What Was the Four-Minute Mile?, 28
    Which Swimming Stroke Is Fastest?, 32
    Why Are Professional Football Players Paid So Much?, 30
    Why Is a Baseball Field Called a Diamond?, 29
    Why Is a Track Field Shaped the Way It Is?, 29
Sprint, 29
Square, 37, 85
Square deal, 86
Square egg, 82
Square number, 53
Square root, 117
Sri Lanka, 41, 101
Star(s), 34-35, 88, 103
Star-clock, 103
Starfish, 87
    Why Do Starfish Have Five Arms?, 87
State lottery, 12
Stevin, Simon, 52
Strategy, 70
Sundial, 71, 103
Sunya, 47
Sumer, 19
Swimming, 32

T

Tacamanaco Indians, 49
Tanzania, 28
Tax, 22, 92
Telephone(s), 108-110
    bill, 94
    book, 66
    numbers, 66, 108, 109, 110
    Can a Computer Remember All the Numbers in a Phone Book?, 66
    Can You Call a Friend Without Using a Phone Number?, 109
    Why Do Telephone Dials Have Both Numbers and Letters?, 109
    Why Do We Need Telephone Numbers?, 108
    Why Does a Phone Bill Have Holes In It?, 94
    Why Do We Need Area Code Numbers?, 110
Tennis, 31
Thaler, 17
Thermometer(s), 48, 80
Thirteen, 9, 11, 91
Thirteen Goddess, 8
Tic-tac-toe machine, 64
Time capsule, 122
    What Is a Time Capsule?, 122
Tire, 83
To, 77
Toes, counting on, 49
Tom Sawyer, 75
Topologist, 119
Track field, 29
Treasuries, 19
Triangle, 85
Triskaidakaphobia, 11
Troy pound, 76
Tum, 77
Turk, The, 67
Twain, Mark, 75

U

Union, 22
Unit price, 90
    What Does Unit Price Mean?, 90
United States, 9, 15, 18, 26, 28, 30, 41, 76, 78, 79, 92, 94, 96, 101, 114
United States Customary System, 78
Universe, 123
Unlucky number, 9
Uphill mile, 122

V

Venice, 19
Vienna, 67
Viete, Francois, 52
Vikings, 75

W

Wages, 25
Wasp(s), 50
Week, 101
Weight, 19, 76, 77, 78
Williams, Ted, 29

Y

Yap, 16
Yard, 78
Yardstick, 74
Year, 100-102, 106

Z

Zero, 45, 47, 48
Zapotec Indian, 8
Zero freaks, 47
ZIP code, 114
    Why Do We Have ZIP Codes?, 114
Zodiac, 34-35